D0082702

Laser Satellite Communications

Morris Katzman, Ed.

Rockwell International

Prentice-Hall, Inc., Englewood Cliffs, NJ 07632

Library of Congress Cataloging-in-Publication Data

Laser satellite communications.

1. Laser communication systems. 2. Astronautics—Optical communication systems. I. Katzman, Morris, 1923–
TK5103.6.L38 1987 621.38′0422 86-20449
ISBN 0-13-523804-8

Cover design: 20/20 Services, Inc.
Manufacturing buyer: S. Gordon Osbourne

*This book was written before Morris Katzman and William Stoelzner joined Rockwell International; therefore, Rockwell accepts no responsibility or liability for its contents.

Printed in the United States of America

10 9 8 7 6 5 4 3 2 1

ISBN 0-13-523804-8 025

Prentice-Hall International (UK) Limited, *London*
Prentice-Hall of Australia Pty. Limited, *Sydney*
Prentice-Hall Canada, *Toronto*
Prentice-Hall Hispanoamericana S.A., *Mexico*
Prentice-Hall of India Private Limited, *New Delhi*
Prentice-Hall of Japan, Inc., *Tokyo*
Prentice-Hall of Southeast Asia Pte. Ltd., *Singapore*
Editora Prentice-Hall do Brasil, Ltda., *Rio de Janeiro*

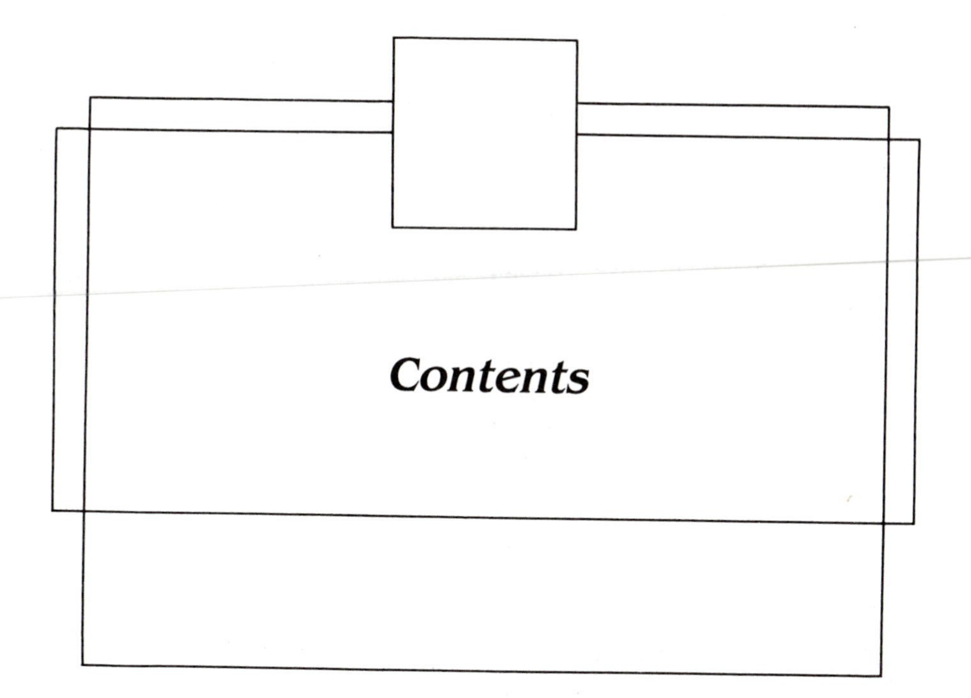

Contents

5 OPTICAL CONFIGURATION AND SYSTEM DESIGN 146

William Stoelzner

6 LASER BEAM POINTING CONTROL, ACQUISITION, AND TRACKING SYSTEMS 190

J. M. Lopez and Dr. K. Yong

A A BRIEF OUTLINE OF AN RF CROSSLINK SYSTEM DESIGN 214

Dr. M. A. King, Jr.

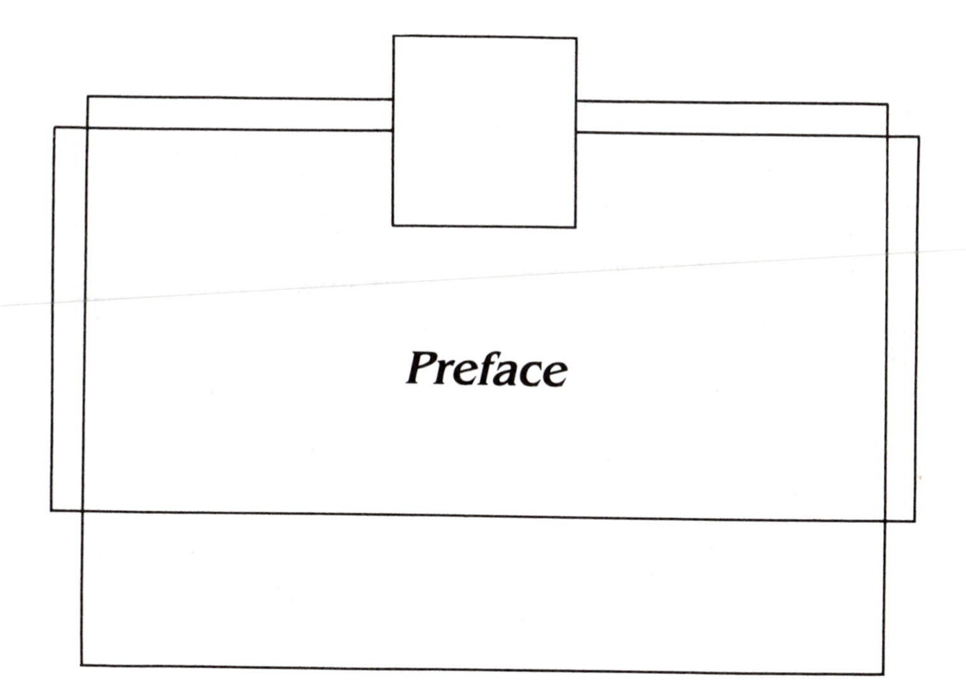

Preface

In 1883, an average family gets into a horse-drawn buggy in Los Angeles and makes the two-day trip to San Diego. In 1983, the great grandchildren get into their family van and make the same trip in two hours. That same year a space shuttle is launched at Cape Canaveral, Florida and travels about the same distance but straight upwards; a ten-minute trip into orbit at an enormous expenditure of energy and cost to the national budget.

In 1883 the typical method of communicating short messages rapidly was via the telegraph. At that time signalling messages at 50 words per minute was adequate to the needs of our average family on their two-day sojourn. At about the same time, Alexander Bell demonstrated our first optical communication system: the photo-phone. Although never more than a laboratory curiosity, the photo-phone represented our desire and need for better and faster communications.

In 1983 our typical family driving between Los Angeles and San Diego turns on the automobile radio and has immediate access to over 75 channels of communications, each of bandwidths up to 40 KHz. The data communications provide current en route weather conditions as viewed from a meteorological satellite, newcasts relayed via satellite links from around the world, or entertainment and recreation. Perhaps, this vehicle might be equipped with a two-way radio or cellular telephone for immediate summoning of help in an emergency or informing parties at either end of their journey as to time of arrival or en route delays. In this one two-hour journey, this family will have utilized more telecommunications service than our family in 1883 would have utilized in its entire lifetime.

Compared to our average family, however, the communications requirements of the space shuttle appear staggering. As the shuttle is launched, on-board computers are linked up with launch control computers providing near instantaneous course and altitude updates as well as relaying back billions of bits of vital health and status parameters of the many on-board shuttle systems. In the two minutes of booster-powered launch, the shuttle will have telecommunicated far more bits of information than were received by our family automobile in its entire two-hour trip. In short, our need for better and faster communications has paralleled and in part has been driven by our mobility. As we reach further out into space and increase our circle of influence on earth, our needs for faster communications will continue to increase.

Even today, less than 20 years after the introduction of satellite relay communications, the system is antiquated and provides limited capability. The popular geo-stationary orbit has become literally cluttered with communications satellites. The earth–satellite–earth relay linking is cumbersome and expensive, requiring complex system networking and expensive remote ground stations.

The alternative is for point-to-point crosslink trunklines between geostationary satellites. However, by international agreement, r.f. crosslinks are restricted due to interference with neighboring satellites. In addition, r.f. packages and antenna sizes have become prohibitive for crosslinks.

Mr. Bell's forgotten invention, the photo-phone, has surfaced again, answering today's communication needs. Only now, instead of modulating the sun, light-wave communications uses the laser. Communications using lasers eliminates the restrictive aspects of satellite crosslinking. The very narrow beamwidths involved (typically $<0.1°$) eliminate interference with neighboring satellites. These narrow beams also provide exceptionally high antenna gains curing the size, weight, and power aspects of the communications terminals. In addition, the optical frequencies can provide essentially unlimited communication bandwidths of greater than 2000 Mbps capability.

Unlike the photo-phone, laser communication is not a laboratory curiosity. A system is currently being built to fly on military satellites in this decade. This system is capable of sending data between hemispheres using just one pair of laser communications transceivers. Although the first of its kind, the development and deployment of this system represents a milestone in the development of a technology that only two decades ago was said to be "a solution in search of a problem." Laser systems are now readily accepted as space-worthy. The design of these systems, as this book will point out, is no longer a mysterious science limited to laboratory and desk-bound academics, but a practical engineering art with technologically well-developed building blocks and analytical tools.

It is the intent of this book to acquaint the reader first with what these system building blocks consist of and some of the unique terminology involved in describing these components. The understanding of how system users and designers allocate performance to the components is crucial. The remainder of the book, provides the reader with some of the considerations that go into the design of these various building blocks. Once acquainted with the technology and terminology, the capabilities of laser satellite communications will appear more attractive and the "mystique" less formidable.

ACKNOWLEDGMENTS

I am greatly indebted to many very special people who made this book possible. I can mention only a few. First, I must thank Robert Howland, President of HDL Communications for launching me on this project and guiding me. I want to thank Louise Welch for her patience in preparing the manuscript. Finally, I must thank my wife Sarah for introducing me to the amazing world of publishing and her constant support and encouragement.

Morris Katzman

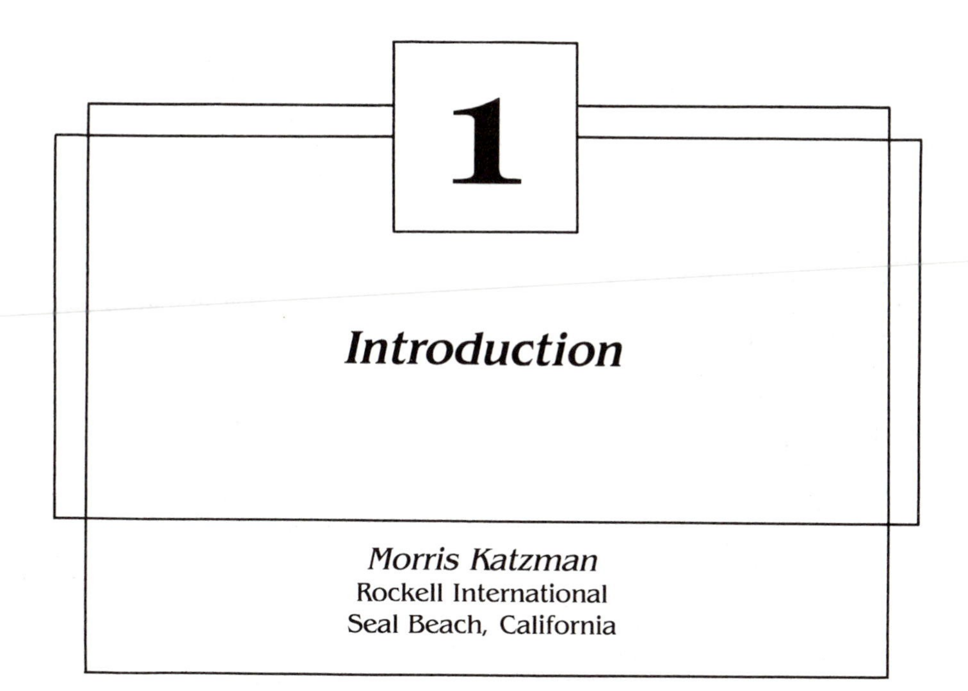

Morris Katzman
Rockell International
Seal Beach, California

1.1 INTRODUCTION TO LASER COMMUNICATIONS (LASERCOM)

The interest in laser communications for satellite applications stems from the much higher operating frequency, some seven or eight orders of magnitude higher than radio frequency (RF) systems. This provides the three main advantages—i.e., greater bandwidth, smaller beam divergence angles, smaller antennas, and new regions of the available spectrum. Information bandwidth, beam divergence angles, and antenna size are all wavelength dependent. RF or microwave wavelengths cover the range from hundreds of meters to less than a centimeter, whereas those of present laser transmitters appropriate for satellite communications vary from less than a micrometer to 10 μm. A comparison of the two types of systems is given in the appendix. A brief comparison of beam divergence angles illustrates the impact of wavelengths.

Consider specifically the comparison between a 3-cm communication system and a 1-μm laser system. The transmitted beam divergence angle varies inversely with aperture diameter and directly with wavelength. Comparing a laser wavelength of 1×10^{-6} m with a microwave value of 3×10^{-2} m and assuming a 3-m antenna for the microwave case and a 10-cm antenna for the laser case, the ratio of solid angles is 10^6. If all other elements of system performance were equal, we would need one millionth the power out at optical wavelengths compared to microwave. (Since this is not so and considering increased space loss, the real ratios are more like two orders of magnitude.) In practice, beam divergence angles go down to microradians. Obviously, this puts a burden on pointing and tracking subsystem designs, but explains the practicality of a laser transmitter with only a fraction of a percent efficiency and a fraction of a watt of transmitter power.

An additional advantage derived from the unique characteristics of laser transmitters is the very narrow (high-peak power) pulses available that make the high data rates practical. Data rates can be as high as multigigabits per second. For example, at 5 gigabits per second about a million telephone channels would be available to provide a kind of "giant trunk line in the sky." In the scientific area, lasercom will play a pivotal role in deep space probe communications.

A simplified functional block diagram of a satellite lasercom system is shown in Figure 1-1. The chapters addressing each block are shown in the figure. Block I, "The Laser," is discussed in detail in Chapter 3, primarily in terms of the AlGaAs diode lasers. Block II, Modulation Techniques, is discussed in both Chapters 2 and 3; Chapter 2, in terms of the Nd;YAG laser and Chapter 3 again in terms of the AlGaAs diode lasers. Blocks III and V, Chapter 5, discuss the optics; Blocks IV and VII, Chapter 6, discuss beam control. Block VI, "Optical Receivers," is discussed in Chapter 4. Chapter 2 provides overall system consideration. It is hoped that by placing this chapter early in the book that the reader will have a better understanding of the details of the subsystems given in the chapters that follow. The communication system described in Chapter 2 is primarily the direct detection system. Heterodyne or homodyne is a feasible approach and R&D efforts are being pursued at present using GaAs diodes. The detection by heterodyne or homodyne techniques is discussed in Chapter 4.

A brief description of the major functional elements is given as an

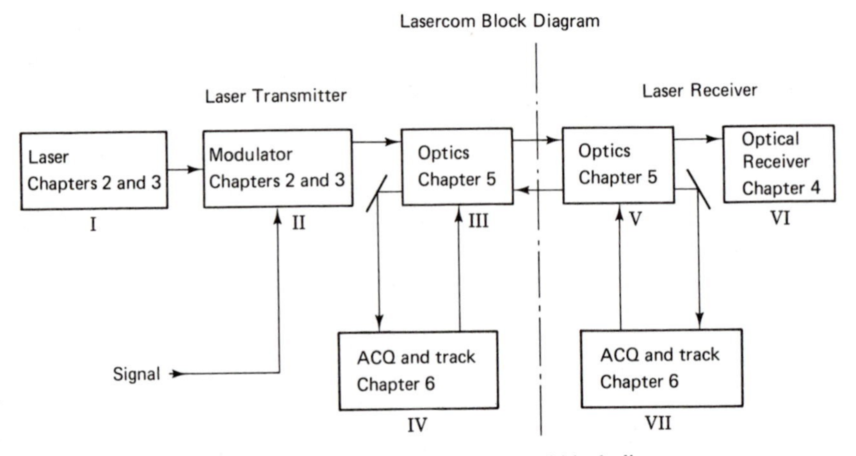

Figure 1-1 Lasercom system functional block diagram.

introduction. There is some overlap in the chapters. The individual authors were allowed this indulgence for the sake of completeness.

1.2 LASER TRANSMITTERS

AlGaAs diode lasers; AlGaAs diode-pumped Nd:YAG; AlGaAs diode-pumped doubled Nd:YAG; and CO_2 are the four principle candidate laser sources. The main characteristics are shown in Table 1-1. The GaAs diodes and Nd:YAG pumped by AlGaAs diodes are the two most practical candidate laser sources at this time and are detailed in Chapters 2 and 3.

The diode laser is small, relatively efficient, rugged, available over a limited wavelength range (approximately 0.8 to 1.7 m), can be directly modulated, and has long potential life ($\sim 10^5$ h). The main disadvantage is the limited power output per diode, so that most applications require the use of diode arrays, leading to beam-combining problems. Another disadvantage is that the projected long life has not been demonstrated. Accelerated life tests are now being conducted. Some shift in output spectrum was observed during an accelerated life test. A shift of about 8 Å was observed after about 600 h at 40°C.

AlGaAs-pumped Nd:YAG avoids beam-combining problems and has a well-developed modulation technique. Sufficient power is usually available to close the link. The small disadvantage is that the well-developed electro-optical modulation equipment is fairly complex. This is essentially the case

TABLE 1-1 CHARACTERISTICS OF PRINCIPAL CANDIDATE LASER SOURCES

Laser Type	*Operating Wavelength*	*Overall Transmitter Efficiency, %*	*Laser Oscillator Efficiency, %*	*Lifetime/ Problem Areas*
AlGaAs	0.89 μm	1.0	5.0 to 10	Laboratory[1] 40,000 h extrapolate 10^5 to 10^6 h.
Nd:YAG	1.06 μm	0.5	0.1 to 7	Essentially established by laser pump diode array, 40,000 h.
Doubled ND:YAG	0.532 μm	0.4	0.08 to 5	Essentially established by laser pump diode array, 40,000 h
CO_2	10.6 μm	0.7[2]	10 to 15	$CO_2 \rightarrow CO$ reaction, seals, and anode and cathode. RF excitation cures many of these problems.

for both fundamental Nd:YAG—i.e., 1.06 μm—and for the doubled Nd:YAG—i.e., 0.532 μm. Diode-pumped Nd:YAG has the diode array lifetime uncertainty problem, as well as wavelength shift with age.

The CO_2 laser is considerably different from the AlGaAs or the Nd:YAG in that it is a gas laser. The two solid-state lasers are electron injected or optically pumped (for the Nd:YAG). The $C0_2$ is pumped by a gas discharge. The significance of this is that CO_2 lasers are essentially discharge tubes with the usual reliability problems introduced by vacuum seals, cathodes, and anodes. To balance the picture, DC-pumped CO_2 lasers have been operated in the laboratory environment for over 10,000 h.[2] As was stated earlier, CO_2 suffers from three deficiencies. First is the lifetime limitations because of the nature of gas discharge devices and the fact that CO_2 chemically reacts and goes to CO with time. Second is the added complexity of heterodyne or homodyne detection because of the requirement of critically aligned local oscillator and signal fields on the mixer crystal surface. Third is the requirement for cryogenically cooled detectors. There are factors that serve to mitigate these effects. The development of waveguide CO_2 lasers replaces the DC gas discharge (and the internal anode and cathode) with an external RF-pumped design. This gives promise of lifetimes in the region of 30,000 h. Some claim is made of the potential of 50,000 h life.[3] At any rate, the design life would have to be demonstrated, and since the laser is not solid state, accelerated life tests cannot be done. The other two factors—i.e., alignment criticality and detector cryogenics—are not considered as difficult as the lifetime factor. Radiative cooling might be adequate since the presently used mercury cadmium telluride detectors constitute a small heat load.

Power-combining technology for GaAs laser diodes will provide a practical way to increase the power and data rates available from laser diodes. Coherent combining using integrated optics technology promises dramatic improvements in diode laser brightness by increasing power in the beam while decreasing beam divergence. In addition to higher available power levels, development is also required for high-data-rate transmitters. Present laser diodes at the required power levels have exhibited nanosecond pulse responses. For multigigabit systems, subnanosecond response is needed. This has been demonstrated at low power levels. Long-term goals would include increased output power, shorter pulse widths, and increased pulse rate to support multigigabit communications.[4,5]

The GaAs semiconductor laser is undergoing continued development. Problems such as failure mechanisms, wavelength shift with age mechanisms, and mode degradation need to be understood so that lifetime can be improved.

Future laser transmitters will make use of such advances in the technology as integrated optics, and monolithically fabricated diode laser array designs. A detailed description of the state of development of diode lasers is given in Chapter 3.

1.2.1 Pulse Modulation Formats of Laser Transmitters

At this time, the most practical modulation scheme for laser communications is the pulse position modulation. AlGaAs diodes require the simplest circuitry to implement pulse modulation. The input current is modulated to generate the required laser pulses. This direct modulation is practical over a very wide frequency range from near dc to 12 GHz, but with low output powers. This area of the technology is treated in great detail in Chapter 3.

For the Nd:YAG lasers, there are basically two control methods that are important for communications—i.e., "Q" switching and cavity dumping. Q switching made possible the development of useful ranging systems despite the extremely low efficiencies of the early ruby laser. It was found that by modulating the "Q" of the resonant cavity, the peak power levels of the pulses were increased by orders of magnitude. This is done by either electro-optically modulating the "Q" or mechanically rotating one mirror of the Fabry-Perot resonator. Modulating the "Q" at the right frequency produces mode locking. The critical frequency is $c/2L$, where c is the speed of light and L is the optical cavity length. Pulses as narrow as 76 ps have been demonstrated, which makes pulse position modulation at multigigabit per second rate practical. This particular capability stems from the dimensional relationship between laser wavelength and resonator. Commonly used laser wavelengths are in the μm region and resonators are in the centimeter region, so that the ratio is about 10^5. Many longitudinal (frequency) modes can be supported and since pulse narrowing varies with the number of modes under the fluorescent band, nano-, and picosecond pulses are available.

Cavity dumping is the opening and closing of a cavity port. One way of accomplishing this is to polarize the radiation in the cavity. The preferred polarization sense can be kept within the cavity, building up the field. Electro-optically rotating the plane of polarization will cause the preferred polarization to be transmitted out of the cavity. The important aspect of cavity dumping for communications is that the pulses can be generated at controllable time locations. Pulse repetition frequencies in the hundreds of kilohertz range are available using cavity dumping.

The modulation format presently being applied is described below. The technique is commonly called *pulse position modulation (PPM)*. PPM employs a synch pulse to define a major time division, often called a "window." The window is then divided into time slots that correspond to the particular data value. Figure 1-2 is an illustration of the PPM format.

Clock resolution and laser pulse-rate capacity are the more important factors that limit data-rate capability of the system. Obviously, the pulse width must not be large compared to the time slot in order to avoid ambiguity. For communications at a gigabit per second data rate, the pulse widths would

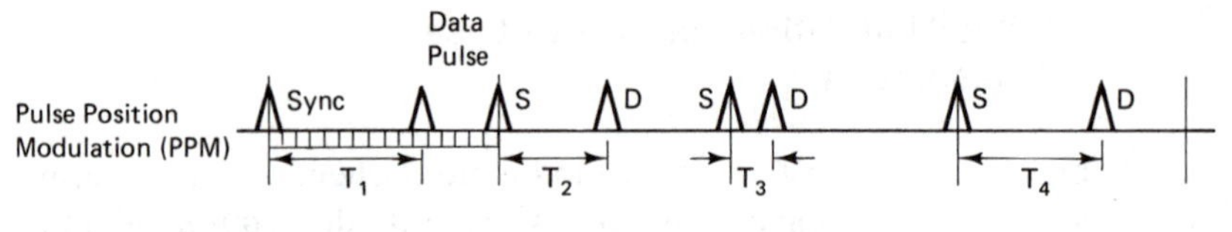

Figure 1-2 Pulse position modulation format.

have to be below 1 ns and the pulse rep rate in the neighborhood of 10^8 pps. This requires mode-locked operation.

In general, it is found that the particular characteristics of laser oscillators can be utilized to provide very effective transmitters for PPM communication systems.

McDonnell-Douglas Corporation has developed a frequency-doubled, mode-locked Nd:YAG laser transmitter.[6] The crystal for doubling is $Ba_2Na(NbO_3)_5$ and is kept at the temperature for phase matching the 1.064 μm fundamental with the 0.532 μm doubled wavelength. The same crystal is acoustically driven to modulate the cavity at the mode-locked frequency. The resulting output with CW pumping is a pulse train of 500-MHz pulse repetition frequency with about one-third ns pulse width. This pulse train, which is plane polarized, is then sent through an electro-optic switch ($LiTaO_2$), which can rotate the plane of polarization by 90° according to the modulation requirements. Rotating the plane of polarization causes the beam to be deflected through a longer path to cause a 1-ns time delay. An unrotated pulse would be undelayed. The two pulse trains are then recombined and the result is modulated pulse train carrying the binary data of 1's and 0's at a 500-MHz data rate and with the two polarization planes at 90° to each other. The additional 500 Mbps is obtained by imposing the additional data with a second modulator similar to the first but on the unrotated pulses, thus obtaining the 1 Gbps data rate total. Modulation formats and techniques are discussed in detail in Chapter 2.

1.3 OPTICAL RECEIVERS

Optical receivers that are described in detail in Chapter 4 can be designed for direct detection, heterodyne detection, or homodyne detection. Direct detection is efficient for cases where large intrinsic gain is available from the detector. Photomultiplier tubes or avalanche photo diodes are examples of high available gain—i.e., a properly designed receiver can provide shot noise or background limited detection. This condition is true for signal photons below 1 μm in wavelength. For longer wavelengths, because of the work functions of the photo emissive surface, the intrinsic gain rapidly falls off. The signal photons simply are too low in energy for the work function energy barrier. From about 1.1 μm to the far infrared, detection usually is thermal-limited, and heterodyne (or homodyne) detection is required.

Heterodyne detection (the mixing of local oscillator beam with the signal beam) provides conversion gain and thus can make the detection process shot noise limited. The heterodyne detection process can be thought of as the mixing of two fields to produce a current proportional to the product. Since one of the fields is the local oscillator field, it can be made large enough so that shot noise produced is larger than the thermal noise in the circuit. For optical heterodyne receivers, CO_2 is the important candidate. The local oscillator is about 92 to 95 percent and the signal is 5 to 8 percent of the total field strength. A difference frequency can be detected by a cooled HgCdTe detector, amplified in a 30-MHz IF amplifier, for example, and detected in a limiter discriminator.

This 30-MHz difference frequency must be maintained within the passband of the IF, making a frequency lock circuit necessary. This can be accomplished by having the resonator frequency of the local oscillator laser controlled by a piezoelectric transducer (PZT) unit. One of the laser resonator mirrors can be mounted on the PZT unit.

The local oscillator can be tuned over a frequency range of 600 MHz through the use of this PZT-controlled mirror. The frequency stabilization is achieved through the use of a Stark cell to control the PZT unit. Basically, the frequency is locked on the Stark absorption line. A dc electric field applied to the cell shifts the frequency of the absorption line and in this way the local oscillator (LO) frequency is tuned. This ability to tune the LO makes laser heterodyne detection possible for space communication where there is relative motion between stations. The resulting doppler frequency shift is sensed and the LO frequency is adjusted to compensate for the doppler shift so as to keep the difference frequency within the passband of the IF amplifier.

As discussed, for heterodyne detection only the frequency relationship must be maintained—the phase relationship need not be maintained. It is interesting to note that in practice both frequency and phase of the LO can be controlled by the PZT controlled mirror. In general, the FM frequency locked receiver is effective up to about 30 Mbps, whereas the phase-locked (homodyne) receiver is effective from 100 Mbps to about 5 Gbps.

CO_2 heterodyne receivers involve certain development risks, as already stated. The detectors must be cooled to about 77°K, the LO and the signal fields must be kept in alignment, and lifetime of the CO_2 discharge tube is considered high risk for the desired long life. Recent advances in waveguide lasers with RF excitation give promise for a three-to five-year life. An effort is underway to explore injection laser heterodyne detection.

Direct detection at 1.06 μm and below in wavelength is more actively being pursued. Basically, it is a pulse threshold detection process. Direct detection receivers have been built and tested.

Detector technology presents one of the more pressing problem areas that must be overcome in order to design multigigabit per second receivers.

The dynamic crossfield photomultiplier (DCFP) was the communications detector for earlier high data-rate laser communication systems. These detectors are bulky and face lifetime problems. Avalanche photodiodes (APD) offer significant space and weight advantages plus reduced power requirements in comparison to DCFPs. However, present silicon APDs are inferior to DCFPs specifically for high-speed applications. On the other hand, III-V alloy heterostructure avalanche photodiodes are attractive detector candidates. The devices offer high-speed response (3.4 ps rise time, FWTM 150 ps), high quantum efficiency (95 percent at 0.53 μm), and small dark currents (3.4×10^{-8}A/cm^2 at one-half breakdown voltage). They are usable in the 0.4 to 1.8 μm wavelength band. At present, these APDs are still in development but space qualifiable devices should be available soon.

A common bane of all avalanche photodiodes is radiation-induced false alarms. Information on the radiation susceptibility of APDs and on ways to mitigate this susceptibility is now being investigated.

1.4 OPTICAL DESIGN CONSIDERATIONS

Optical design options, which are covered in Chapter 5, are severely constrained. The basic choices in optics design are (1) reflective or refractive optics, and (2) gimballed mirror flat or gimballed telescope for beam steering. Refractive optics offer a simpler, well-developed technology, but a weight penalty is extracted so that presently the preferred direction is toward reflective optics.

Cassegrain optics often is used and a single telescope can be used for both transmitter and receiver. One way to perform the diplexing function is through the use of circularly polarized beams. Each lasercom can transmit one sense of polarization and receive the other polarization. A circularly polarized beam when passed through a quarter-wave plate changes to a linearly polarized beam. A half-wave plate rotates the plane of polarization by 90°. The plane of polarization in the transmit optics path is at a 90° angle from the plane of polarization of the receiver optics path. By insertion or withdrawal of the half-wave plate, the receive and transmit polarization planes can be reversed. Using these techniques, one station can transmit right-hand polarization and receive left-hand polarization, and the other can transmit left-hand polarization and receive right-hand polarization.

Figure 1-3 shows a basic functional optical block diagram of such a station showing the common optics, diplexer, transmitter optics, and receiver optics.[6] Notice the two-stage detector array for acquisition and track. This is a generalized representation of a multistage acquisition and track subsystem.

This introductory chapter only touched upon the main functional ele-

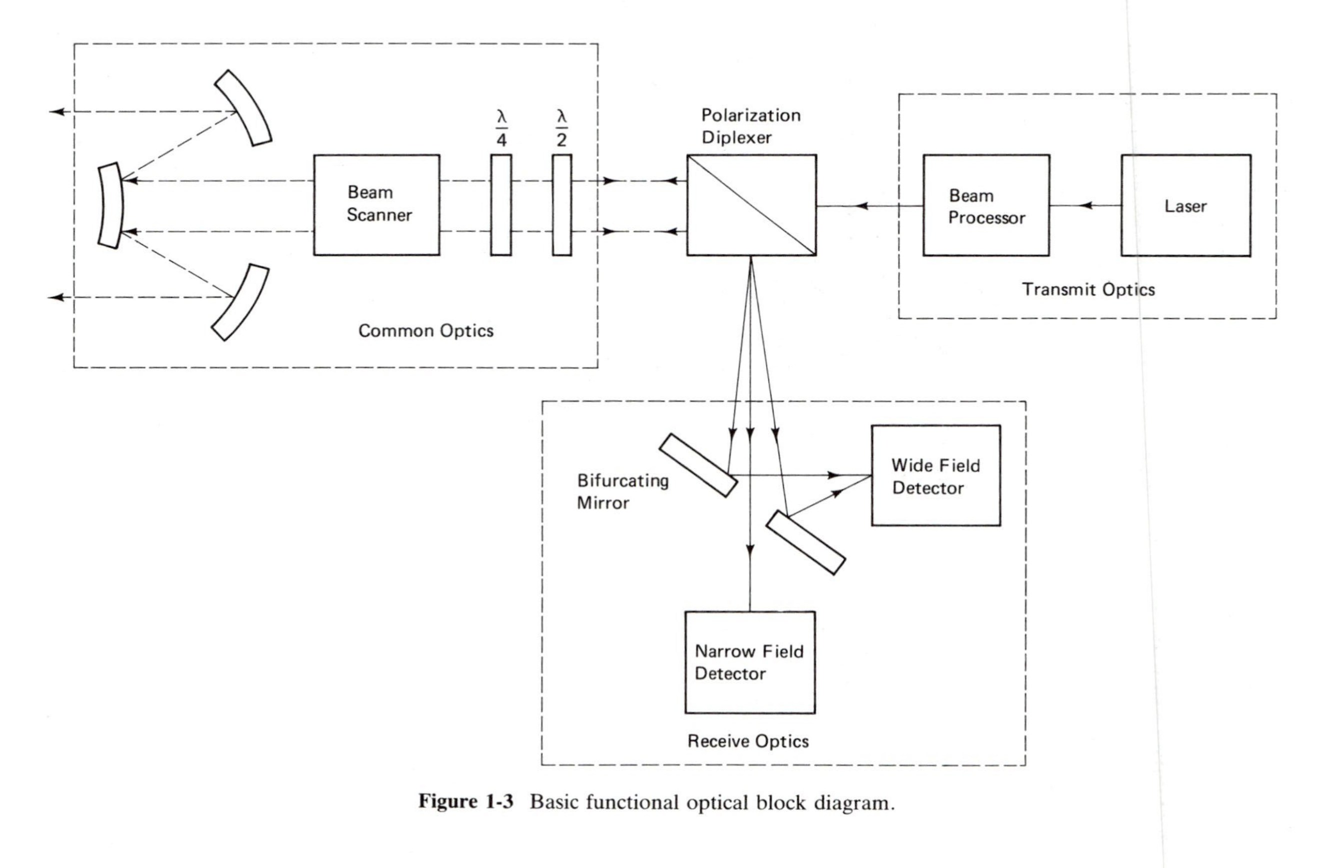

Figure 1-3 Basic functional optical block diagram.

ments of the entire system. The following chapters describe in detail the various aspects of laser communications in space. It is hoped that by placing an overall system chapter (Chapter 2) early in the book, the reader will be motivated and will have a greater appreciation for the details that follow in succeeding chapters. Understanding the function and performance impact of the many design aspects and building blocks of the entire system should provide a degree of coherence to a fairly complex picture. Additionally, a few functional elements are detailed in the system chapter simply because they are not included in any of the succeeding chapters and are necessary to complete the overall presentation.

REFERENCES

1. M. Ettenberg and H. Kressel, "The Reliability of Al(Ga)As CW Laser Diodes." *IEEE J. of Quantum Electronics*, Vol. QE-16, p. 186, Feb. 1980.
2. J. H. McElroy et al., "CO_2 Laser Communication Systems for Near Earth Space Applications," *Proc. IEEE*, Vol. 65, p. 246 (Feb. 1977).
3. J. Peuso, J. J. Degnan, and E. Hochuli, "Life Test Results for an Ensemble of CO_2 Lasers", TM 79536, p. 1, NASA, Goddard Space Flight Center, Greenbelt, MD Apr. 1970.
4. G. A. Evans et al., "Progress Toward a Monolithically Integrated Coherent Diode Laser Array," SD-TR-81-7, The Aerospace Corporation, El Segundo, CA. July 1981.
5. Laser Diode ML-2000, *Mitsubishi Technical Note*, p. 10, Dec. 1978.
6. M. Ross et al., "Space Optical Communications with the Nd:YAG Laser," *Proc. IEEE*, Vol. 66, pp. 327, 328 (Mar. 1978).

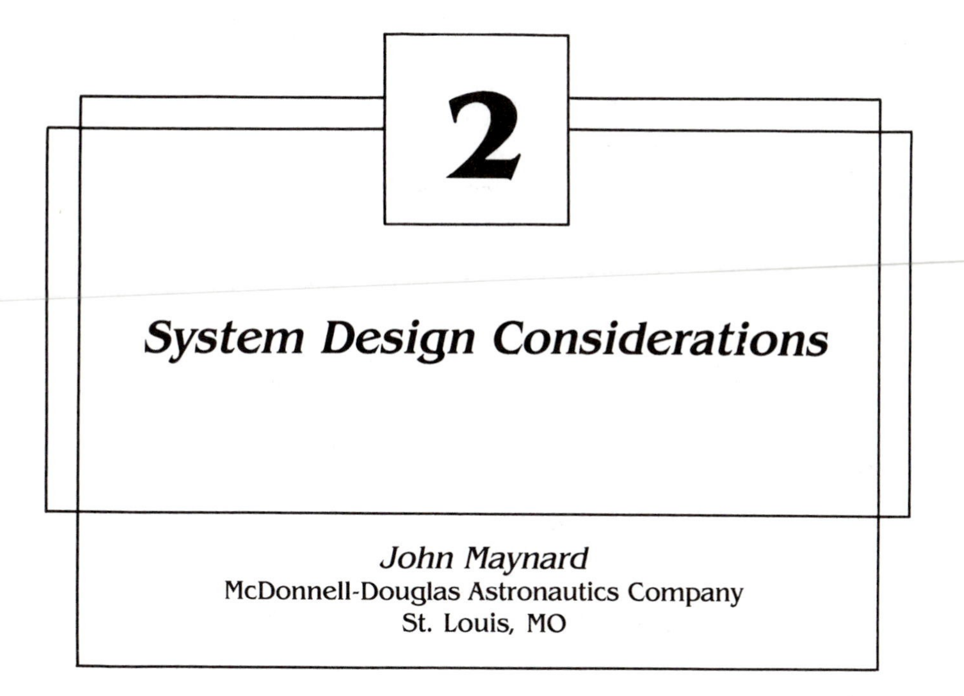

2

System Design Considerations

John Maynard
McDonnell-Douglas Astronautics Company
St. Louis, MO

2.1 INTRODUCTION

The key blocks of a satellite-borne laser communications system are very much analogous to those of RF or microwave systems. In general, a laser communications transmitter system consists of data-handling electronics, a modulated carrier source, a transmission system (optics), and an antenna (usually a telescope). The laser communications receiver is likewise similar to an RF system in fundamental functions except that a photodetector is used as the actual energy-detecting element. For both types of systems, control of operating modes and pointing control is handled with a control computer. Figure 2-1 shows the typical interrelationship of these elements in a laser communications system.

In general, many of the techniques used in evaluating RF communications systems are directly applicable toward evaluating a laser communications system. Since early workers in the optical communications field were experienced in communications principles and techniques, much of the terminology used in describing laser communications has been adapted from its RF counterparts. Concepts such as antenna gain and antenna efficiency, which generally would not be used to characterize an optical system, have been adapted in describing laser communications systems in place of more generally associated "optical" terms of beam waist size and wavefront quality. Thus, workers approaching the laser communications field from an optical back-

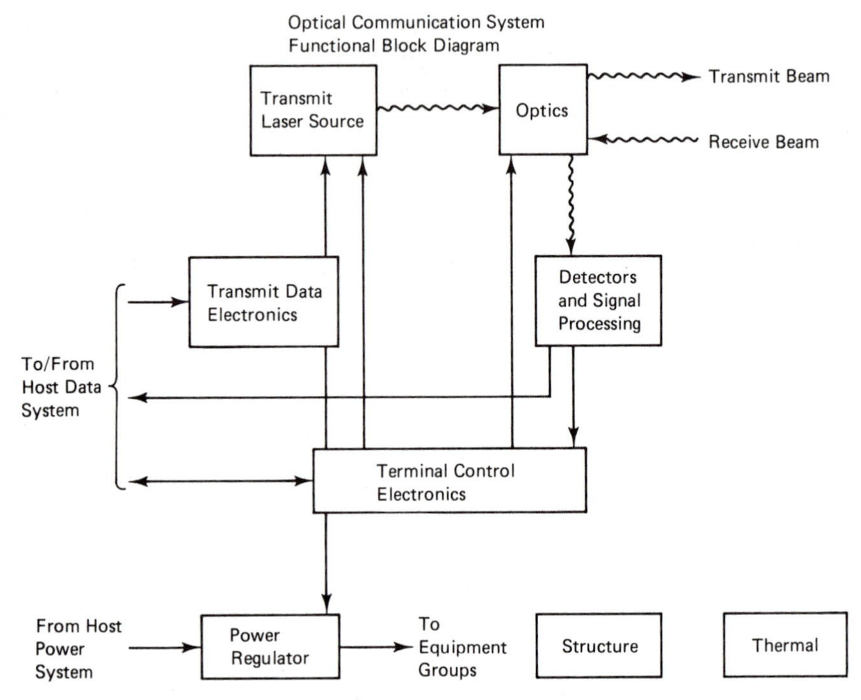

Figure 2-1 Optical communications system block diagram. Optical communications transceiver performs similar functions to RF system. The optics block includes both feed optics and the telescope (antenna). Optical communications systems almost always include cooperative optical tracking for precision pointing of narrow beams and receiver viewfields.

ground must first familiarize themselves with the applied terminology. Conversely, workers approaching optical communications from an RF communications background must understand the application of optical principles to conventional terminology.

2.2 THE LINK EQUATION

For both optical and RF communications links, system parameters and performance margins usually are established by means of the *link equation*. This equation provides a means of calculating the detected energy at the receiver based on modulated carrier power, system gains and losses, and effects of beam propagation through the transmitting medium. Performance evaluation is made by determining system probability of error, P_E, for a detected signal to noise ratio, *SNR*. A performance margin particularly for digital links often is established by determining the ratio of detected energy to the energy required to establish a minimum performance criteria.

For a typical communications link as illustrated in Figure 2-2, there are numerous parameters that govern the performance of the system. These parameters represent specific transmit and receive terminal operating characteristics or physical characteristics associated with the propagation of the transmitted energy. The link equation is the closed form expression that is used to predict link performance. This equation most often is used in the logarithmic form, and usually is written to predict signal margin at the receiver:[1]

$$M(\text{dB}) = 10\log(P) - L_t + G_T - L_p - L_n - L - L_{\text{LINK}} + G_R - L_R - 10\log(QE) - L_{\text{proc}} - 10\log(S_{\text{req}}) \qquad (2\text{-}1)$$

where

M	=	ratio (dB) of detected signal to required signal to establish minimum performance requirements
P	=	emitting source output power (W)
L_t	=	transmitter system antenna feed losses (dB)
G_T	=	transmit antenna (telescope) gain (dB)
L_n	=	transmit antenna wavefront efficiency (dB)
L_p	=	transmit beam pointing loss (dB)
L	=	free space propagation loss (dB)

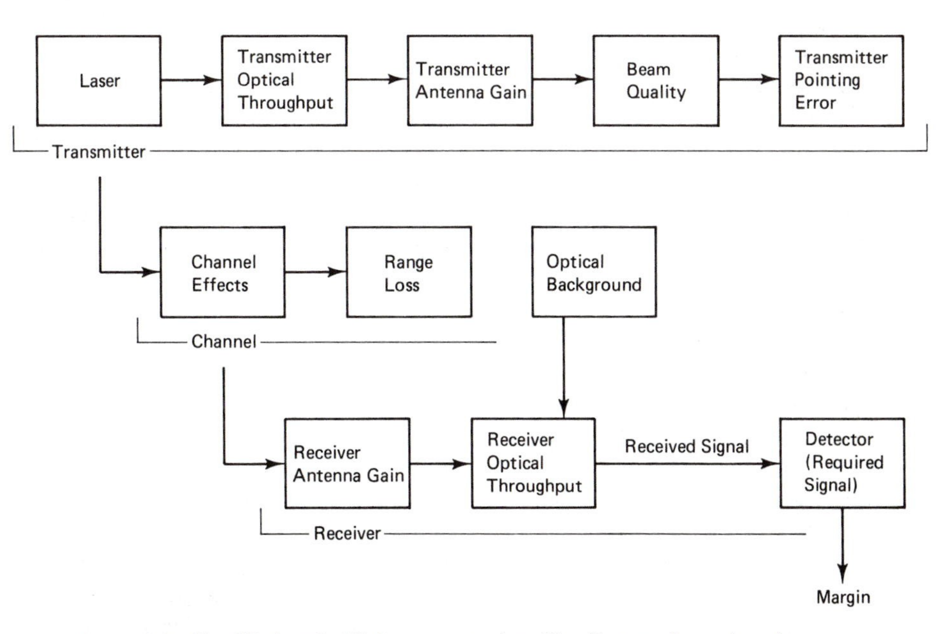

Figure 2-2 Simplified optical link representation. The diagram shows the relationship between the various elements of the link equation. The system designer must perform the parametric trade-off to determine the optimum system design.

L_{LINK} = additional propagation losses such as polarization loss, and medium attenuation (dB)
G_R = receive antenna gain (dB)
L_R = receive antenna signal relay losses (dB)
QE = detector energy detection efficiency
S_{req} = signal required for minimum performance based on link noise background, and detector noise.
L_{proc} = data handling electronics processing losses.

2.3 THE TRANSMITTER TERMINAL

Laser Sources

Although Chapter 3 treats laser diode sources, a brief review of available laser sources is given for completeness. There are a multitude of laser types usable as sources for laser communications systems. Particular laser types are generally identified by the type of material that is used as the "gain medium." Of the five general categories of laser types, three are most often considered for use in communications systems: gas lasers, solid-state lasers, and semiconductor lasers. Selection of which laser family is to be used for a particular communications system is dependent upon a number of factors including link range, propagation medium, data rate, and platform limitations.

All lasers, no matter what type, operate on the same principles.[2] A laser consists of an extremely high-Q cavity resonator built around an energy amplifier (Figure 2-3). Amplification is achieved by exciting or "pumping" the medium to a higher level metastable energy state.

The method of pumping a laser depends upon the type of amplification medium. In the case of gas lasers, the usual source of pumping is by an electric discharge through the gas. Two of the most common types of gas lasers are the helium-neon laser, HeNe, and the carbon dioxide, CO_2, laser. Although the helium-neon laser with its familiar deep red-colored light is by far the laser of most common usage because of its manufacturability and low user maintenance, the CO_2 laser is the gas laser most often selected for communications systems. The reasons for the dominance of the CO_2 laser over the HeNe for communications applications are total output power capability and overall efficiency. The CO_2 laser, which emits radiation in the far infrared (10.6 μm), is the most efficient of all lasers and can produce average power outputs of several kilowatts.

In the CO_2 laser, electric discharge excites the CO_2 molecules to a high-order vibrational quasistable mode that serves as the energy state for the inverted population of the gain medium. Transition from this vibrational mode to the base mode releases energy in the 10.6-μm wavelength band. The line width of the energy transition is broad, permitting the CO_2 laser to be "tuned"

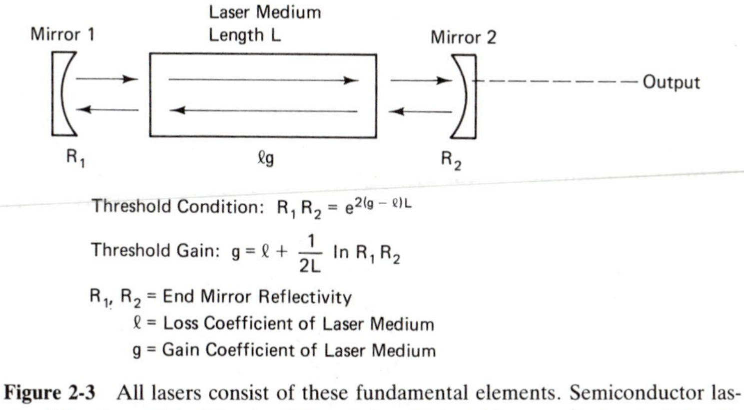

Figure 2-3 All lasers consist of these fundamental elements. Semiconductor lasers utilize the polished facets of the semiconductor chip as end mirrors. Like all oscillators, lasing will commence (threshold) when optical gain just equals roundtrip optical loss.

over a range from around 10.4 μm to 10.6 μm. Tuning is typically accomplished by the use of a diffraction grating in the cavity, which spoils the Q of the cavity for all but the desired wavelength.

Due to the long wavelengths of the CO_2 laser, the laser "optics" generally are made of a metal such as selenium or germanium, which is easily machined. One of the advantages of a CO_2 laser is that for the 10.6-μm wavelength, good "optical-quality" surfaces require surface finishes on the order of 1 μm, which can be achieved and measured with conventional machining technology. The long wavelength also makes CO_2 systems amenable to heterodyne modulation formats further improving link efficiency.[3]

Despite the acknowledged attributes of the CO_2 laser, the laser type that to date has gained the widest acceptance as providing the most viable technology for long-range space communications is the solid-state laser. The most conventional solid-state laser is the Neodymium:YAG (yttrium, aluminum, garnet). In this laser, which is similar to the ruby laser, the light amplification medium is a rod made of a crystalline material (YAG) lightly doped with neodymium. Optical energy is used to "pump" the neodymium ions to a metastable energy state. Decay of these atoms emits light with a fundamental wavelength of 1.064 μm. Historically, for the ruby laser, a conventional photographic flash tube was used to pump the rod. Since the early days of the ruby laser, a significant amount of development effort has taken place to improve pump sources. Today, laboratory versions of the Nd:YAG laser are pumped either with high-intensity tungsten filament lamps or continuously operating ion arc lamps. Although these lasers can produce extremely high peak output powers, laboratory versions are not particularly efficient in terms of electrical pump energy to output energy. A recently developed version of the Nd:YAG laser achieves the efficiencies necessary for a spaceborne laser communications system by utilizing arrays of semiconductor laser diodes to pump the Nd:YAG rod. Since the diode output can be thermally tuned to match the Nd:YAG absorption bands precisely, this laser can be relatively efficient, with efficiencies >7% reported.[4]

In addition to YAG other types of crystalline host materials for the Nd ion have been explored. These include glass, yttrium aluminum oxide, or yttrium lithium fluoride. Although these host materials all lase at slightly different wavelengths, the principles of operation are the same. One additional variation of the neodymium laser is the use of an intracavity element known as a *frequency doubler*. This element, which is an optical crystal usually of lithium niobate ($LiNbO_3$), barium sodium niobate ($Ba_2NaNb_5O_{15}$), or quartz, has the property that at a certain temperature, known as the phase match temperature, a portion of the incident energy is translated to its second harmonic. This process, referred to as second harmonic generation (SHG), is particularly efficient when the frequency doubling crystal is placed within the laser cavity.[5] For neodymium-doped laser materials with fundamental wavelengths of 1.06 μm, the second harmonic energy is 0.532 μm. The deep green light is useful for applications requiring visible illumination and falls in a wavelength regime where photodetectors generally are most sensitive.

2.3.1 Modulators

Modulation techniques for laser communications systems are as varied as the laser themselves. The simplest modulation technique is to vary the electrical pump energy being used to drive the laser. This technique is useful in areas where the pump mechanism is responsive to the frequencies of interest, and the laser cavities themselves can be effectively controlled. In practice, the only laser type in which this approach is meaningful is the semiconductor laser diode, and it will be described in Chapter 3.

All other laser types employ a separate optical element that may be located either within or external to the optical cavity of the laser to modulate the light output. Birefringence modulators, which rotate the plane of polarization of the incident light, are the most common type of optical modulator. These devices utilize the electro-optical properties of crystalline material to modulate the laser output.[6]

The mechanism for electro-optical modulation makes use of electric-field-induced birefringence. Linearly polarized light can be described as the vector sum of two counter-rotating circularly polarized light vectors, shown in Figure 2-4.

OA and **OB** are the two counter-rotating vectors representing right- and left-handed circularly polarized radiation. As can be seen, going from $\mathbf{T_1}$ to $\mathbf{T_5}$ the two rotation vectors add up to the vertical oscillating vector **OC** representing linearly polarized radiation. This forms the essential element in modulation through electric-field-induced birefringence. Modulation is accomplished as follows. Consider, for example, rotating vector **OB**. Let's as-

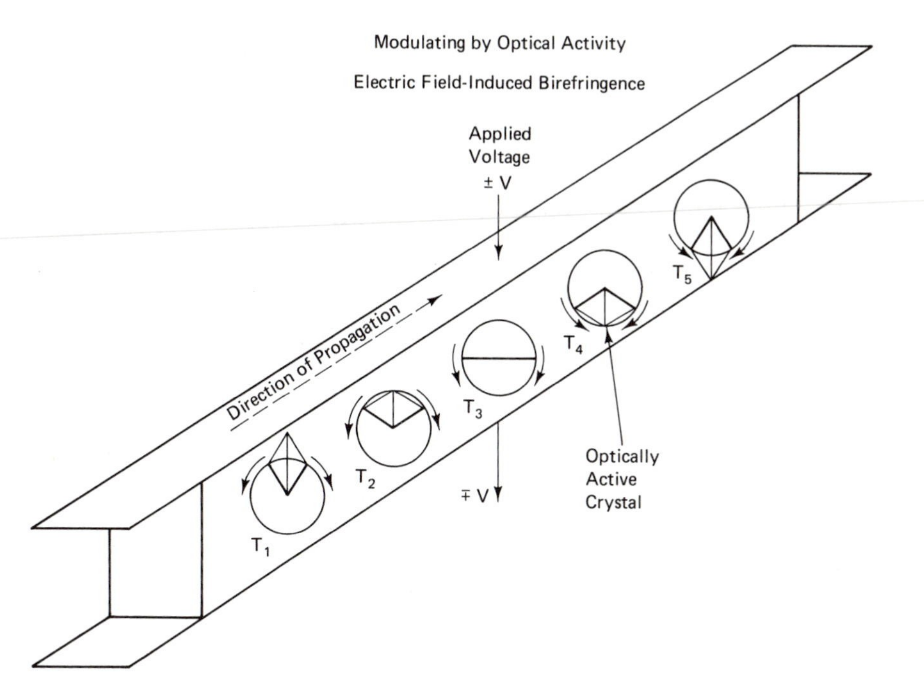

Figure 2-4 Linearly polarized light traveling through modulator crystal may be thought of as resultant of two counter rotating vectors of circularly polarized light. With no voltage applied, linear vector exits in the same plane of orientation. The figure shows the vectors as they would exist in the modulating crystal with the voltage off—no rotating of the plane of polarization.

sume that the induced birefringence causes a reduction in velocity through crystalline medium of vector **OB** whereas vector **OA** retains its original velocity. The vector seen at $\mathbf{T_5}$ will no longer be vertical but will have been rotated by some angle, which is a function of the applied voltage as shown in Figure 2-5.

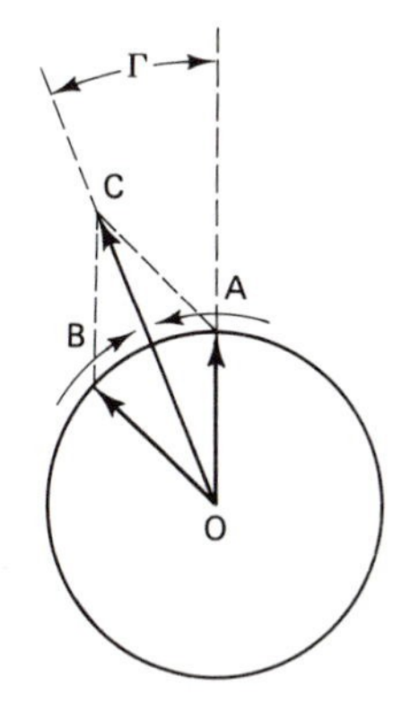

Figure 2-5 Applying an electric field across the modulator causes the retardation of one of the counter-rotating circularly polarized vectors. The resultant at modulator output is a linearly polarized light of rotated vector.

The reduced velocity for the right-handed circularly polarized radiation is called the retardation, and crystals that exhibit this effect are called "optically active."

The field-induced birefringence Δn can be expressed in the form

$$\Delta n = n_0^3 \gamma E = n_0^3 \gamma \frac{V}{d} \tag{2-2}$$

where

n_o = the unperturbed refractive index
γ = an appropriate electro-optic coefficient
E = applied electric field
d = gap over which the electric field is applied
V = applied voltage

After traversing a length L of the optically active crystal, a single optical beam, initially plane-polarized will emerge rotated through an angle Γ, which is dependent upon the applied voltage V and the crystal length L. The voltage is applied across the crystal width d and the resultant rotation can be expressed as:

$$\Gamma = \frac{2\pi L}{\lambda_0} \Delta n \text{ rad} \tag{2-3}$$

Where λ_0 is the free space wavelength.

If we define a specific voltage

$$V_\pi = \frac{\lambda o}{2n_o^3 \gamma}$$

and substitute this into the foregoing expression for Γ and substitute the expression given for Δn, we get the simplified equation:

$$\Gamma = \frac{V}{V_\pi} \cdot \frac{L}{d} \cdot \pi \text{ rad} \tag{2-4}$$

The rotation angle Γ often is called the retardation.

This final expression for the induced retardation shows that to maximize Γ, d must be minimized and L maximized.

Cavity dumping.

One way of employing optical activity for laser modulation is by cavity dumping.[7,8] The optical and electro-optical elements of the technique are shown in Figure 2-6.

The laser resonator illustrated is an L-shaped cavity bounded by the two mirrors labeled "resonator end mirrors." These mirrors are coated for maximum reflectance for 1.06 μm wavelength radiation. Between pulses, the oscillating photon field is built up by the Nd:YAG rod, which is continuously pumped. The preferred polarization is the one with the electric vector in the plane of the diagonal surface of the polarizing beam splitter inside the cube as shown. This is the polarization that will be reflected back and forth between the two resonator end mirrors. In the presence of optical gain in the Nd:YAG crystal, the photon field builds up to a given value. When an output pulse is required, a specific value of electric field is applied to the modulator crystal. This causes a 90° rotation of the plane of polarization. The diagonal surface now becomes transmitting (instead of reflective) and an output pulse leaves the laser as shown at the left side of the figure, thereby depleting the gain. In this way, nanosecond pulses can be generated in a highly controlled way in the time domain satisfying the communications requirements of pulse position modulation.

A complete discussion of the quantum physics involved in the pumping and spontaneous and induced emission is beyond the scope of this book, but a qualitative description will be instructive. It was mentioned before that pumping the Nd:YAG laser rod is done either by white light from a gas

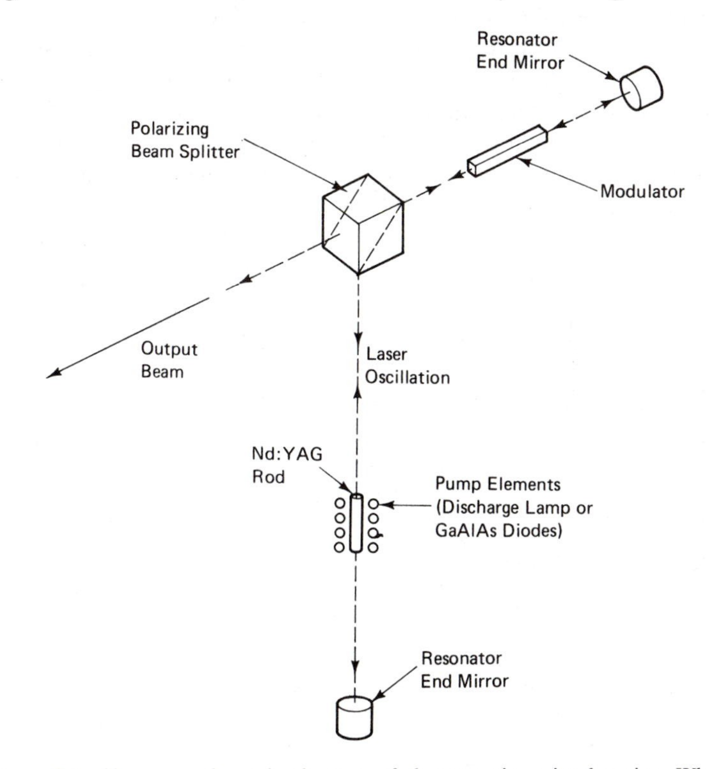

Figure 2-6 Shown are the main elements of electro-optic cavity dumping. When the polarization is aligned so that the *E* vector is in the plane of the diagonal surface of the polarizing beam splitter, the radiation is reflected and stays inside of the laser cavity. When the voltage is applied to the modulator, the polarization is rotated by 90° and the diagonal surface becomes transparent to the radiation and the laser pulse exits to the left as shown.

discharge lamp or by the laser radiation from GaAlAs diode lasers. The pumping action raises electrons from a lower to a higher energy state. In this way, energy is stored in the crystal. A photon traveling through such a medium would trigger electrons to drop to the lower energy state emitting radiant energy thereby adding to the photon field strength. In effect, the traveling photons would experience gain. As long as the loop gain is greater than unity, the photon field strength would increase. In this way, the energy that was first stored in the electrons of the Nd^{3+} atom is now stored in the photon field. Cavity dumping makes use of the energy stored in the photon field and switches open the output port at the proper time, thereby releasing a pulse of laser radiation. Energy can be retained in the energy states of the electrons by using Q-switching. In the cavity dumping mode, the end mirrors making up the laser resonator cavity are highly reflecting. If we were to replace the mirrors with nonreflecting surfaces as shown in Figure 2-6—e.g., highly transmitting or absorbing surfaces (low "Q")—the photon field would not build up since photons would not be reflected back and forth inside the laser crystal and so would not cause the triggering of higher energy electrons to give up their stored energy to the photon field. Continued pumping would increase the electrons stored energy up to some specific level depending upon electron lifetime in the upper state. Now when an output pulse is required the back mirror is made highly reflecting and the output mirror is made partly transparent (high "Q"). The photon field builds up very rapidly (the time that it takes for photons to make a few roundtrips within the crystal) and a pulse

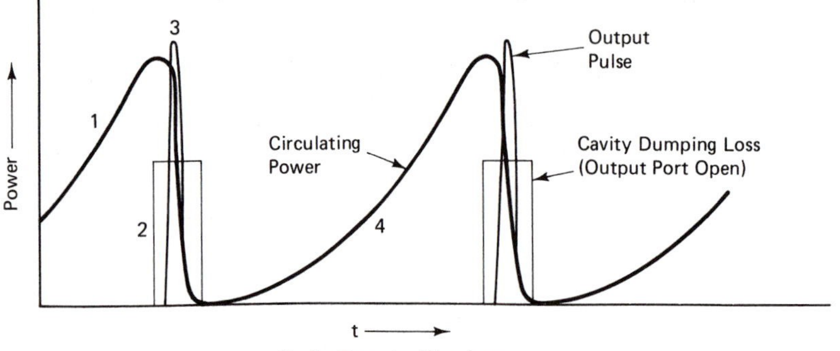

Cavity Dumping Waveforms

Operating Sequence

1. Circulating power builds up to peak
2. High coupling loss switched on (the output port is opened)
3. Power is dumped quickly from cavity
4. Circulating power builds up again

Figure 2-7 With no voltage applied to the modulator pump in a cavity dumped laser, energy is stored in oscillating field between resonator end mirrors. By rapidly applying a voltage to the modulator, the polarization vector of the circulating energy will switch to permit rapid output dumping of all stored optical energy.

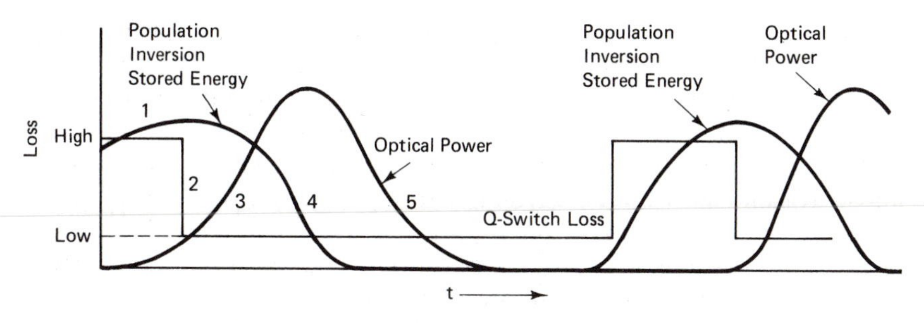

Operating Sequence

1. Gain medium excited with Q-switch in high loss state (population inversion build up)
2. Q-switch switched to low loss state (high Q)
3. Power in resonator builds up exponentially
4. Stored energy is extracted, gain drops to zero (laser pulse is emitted)
5. Power in resonator decays exponentially

Figure 2-8 Waveforms operating in the Q-switched mode—with voltage applied to the modulator, lasing is suppressed due to high resonator cavity loss (low "Q" state). Rod pump energy is used to charge up rod to extremely high gain state. Sudden transition to high-cavity "Q" (by removing modulator voltage) permits spontaneous transfer of energy to circulating power in cavity. Output power is a fixed percentage of circulating power. Lasing continues until gain is fully depleted.

of laser radiation leaves the cavity. The pulse duration typically is in the nanosecond range for both cavity dumping and Q-switching.

For the cavity dumping mode and the Q-switched mode, the respective wave forms are shown in Figures 2-7 and 2-8. In practice, the cavity dumping technique provides higher repetition frequencies, up to about 400 kpps, and the "Q" switching provides higher peak power but at lower repetition frequencies, up to about 14 kpps.

Some of the practical aspects of this kind of modulation would include the cavity losses introduced and the required power. Since the modulator is located inside of the laser cavity, the regeneration will multiply whatever loss exists in the modulating crystal and polarizing beam-splitter cube. This will add to the system power requirements. There also is some power consumed in providing the modulating voltage across the crystal. This will vary according to the modulation rate (the required system bit rate). The power consumed will vary typically from a few watts for megabit rates to tens of watts for gigabit rates.

2.4 ANTENNAS

Both RF and laser communications systems utilize the propagation of electromagnetic energy to transmit information across the link. It is reassuring to note that both system types are governed by the same basic principles of electromagnetics—i.e., Maxwell's equations.[9] However, the applied terminology differs depending upon whether one approaches the subject from an RF or optical background. Both systems utilize antennas to transmit or collect the electromagnetic energy. The purpose of the antenna is to direct the transmitted energy to the receiver or to focus the receiving detector on the transmit terminal. RF antennas can be anything between a simple dipole (such as a common television antenna) or a high-gain Cassegrainian dish (typically seen at satellite ground stations). The geometry of the antenna depends on the frequency of the carrier and the desired far-field pattern.

Laser communications systems also utilize antennas to direct the transmitted energy. For optical systems, however, these antennas are nothing more than conventional design telescopes, where the size and geometry are dictated by the wavelength and system requirements.

At this point, it is worth reviewing the concept and mathematics of antenna patterns and far field gain. The transmit gain of an antenna as applied to optical systems is the ratio of the radiation intensity from the antenna, to the radiation intensity of an ideal isotropic radiator driven by the same input power.[10] The shape of the antenna pattern is predicted using the principles and mathematics of diffraction. Specifically, the distribution of energy from an antenna is predicted by conventional diffraction integrals.[11]

An inherent property of most lasers used in communications systems for space is that the output beams are circularly symmetric and have E-field amplitudes that are Gaussian distributed. When these factors are incorporated into the diffraction integral, the far-field pattern of an optical antenna can be expressed as the equation:

$$\begin{aligned} G(\theta) &= \frac{\bar{E}(k\theta,L)}{\frac{1}{4\pi}} \\ &= \frac{8\pi^2}{\lambda L}\, ie^{-ikL(1+\theta^2)} \int_0^\infty \bar{E}(r_1)e^{-\,ikr_1^2/L}\, J_0\,(k\theta r_1)r_1 dr_1 \end{aligned} \tag{2-5}$$

where

$G(\mathrm{T})$ = far field gain at off-axis angle θ
L = range distance
k = $2\pi/\lambda$
λ = wavelength
$\bar{E}\,(k\Theta,L)$ = electric vector amplitude in far field
$\bar{E}\,(r_1)$ = electric vector amplitude in plane of transmit aperture
r_1 = radius of transmit aperture

Two common antenna geometries are: the unobscured untruncated telescope (typical of most refracting telescopes), and the typical Cassegrainian telescope with central obscuration radius a and truncation radius b (shown in Figure 2-9) and discussed in greater detail in Chapter 5, Sec. 5.7.2. For the first geometry on axis, Equation (2–5) reduces to the equation:

$$G_{dB} = 10 \log_{10} G(o) = 10 \log_{10} \frac{32}{\theta_t^2} \tag{2-6}$$

where

G_{dB} = on-axis gain in decibels
$G(0)$ = evaluation of (2-5) at $\theta = 0$
θ_t = full width half maximum beam divergence angle

For the Cassegrainian dish, the on-axis gain becomes:

$$G_{dB} = 10 \log_{10} G(o) = 10 \log_{10} \left[2 \left(\frac{2\pi r_o}{\lambda} \right)^2 \left(e^{-b^2/r_o^2} - e^{-a^2/r_o^2} \right)^2 \right] \tag{2-7}$$

where

r_o = $1/e^2$ beam radius at the telescope aperture,
a = obscuration radius
b = truncation radius

2.4.1 Antenna Parametrics

Chapter 5 discusses the optical design issues of telescope antenna in detail. In addition, several papers have been written describing the proper selection of telescope parameters to meet system requirements.[12] Some key relationships, however, are worth reviewing at the start of system definition to establish nominal operating parameters. One of those relationships is seen in Figure 2-10, where the far-field pattern for two telescope geometries has been graphed. From Equations (2-6) and (2-7), the relationship between beamwidth and far-field gain are evident. Also from these equations, it can be seen that the gain varies inversely as the square of the wavelength giving rise to the fundamental advantage of laser communications: extremely high antenna gains for relatively small antenna diameters. Finally, Figures 2-10 and 2-11 show the effects of obscuration, truncation, and beam radius for a Cassegrainian system.

2.4.2 Transmitter Antenna Efficiency

For RF antennas, the primary consideration for efficiency is the proper impedance matching to couple as much energy as possible to the free space and reflect a minimal amount of energy back to the carrier source. In optical

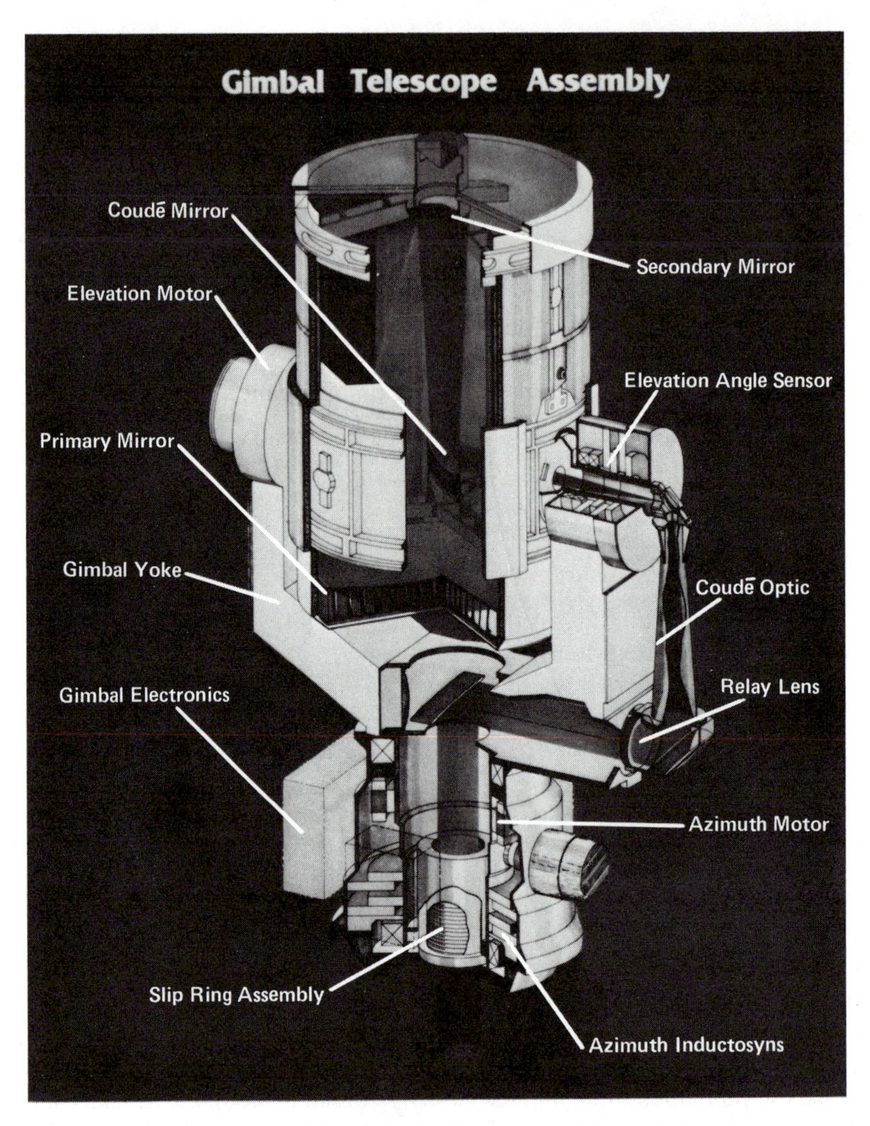

Figure 2-9 Illustration shows the construction of a typical spaceborne optical communications telescope designed by Eastman Kodak. For the unit shown, coarse system pointing is achieved by gimballing the entire telescope barrel. This telescope is 190-mm diameter with a central obstruction of < 50 mm diameter. The optimum beam diameter is ~ 150 mm. The thermal hardware enables the telescope to maintain diffraction limited performance in the presence of space thermal environments. The telescope alone weighs less than 9 lb. (Courtesy McDonnell-Douglas)

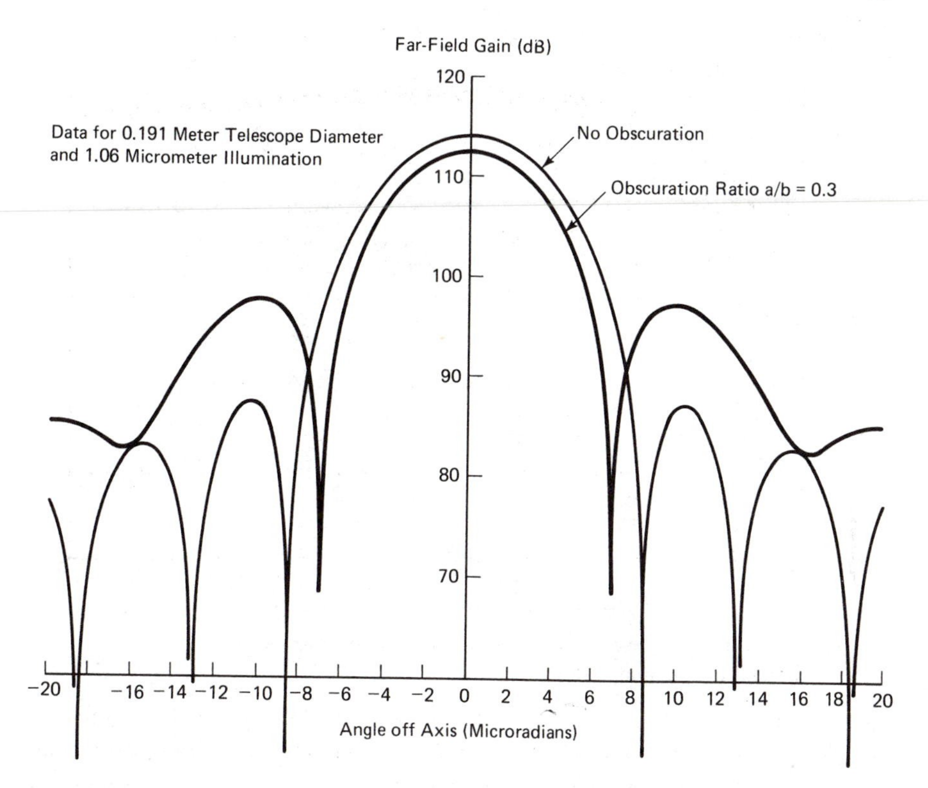

Figure 2-10 Typical antenna far-field patterns shown are the far-field antenna patterns for a 191-mm telescope with and without a central obscuration. Note the effect of broadening in the sidelobes due to the obscuration. In this example, the beam size has been optimized to the telescope based on operation of 1.06-μm wavelength. Usual system operation is for beam pointing to stay within the e^{-1} points of the central lobe.

systems, impedance matching is achieved by the use of precision optical coatings on the surfaces at the individual elements (mirrors and lenses).

Due to the extremely short wavelengths involved, the efficiency of optical antenna usually is limited by the ability to control the physical contour of the individual reflecting or transmitting surfaces. For RF wavelengths of several millimeters controlling physical tolerances of the antenna surfaces to less than 1 percent of the wavelength usually presents no great difficulty. However, for optical systems with wavelengths on the order of 1 μm, mechanical tolerances become extremely tight. So much so, in fact, that the amount of wavefront phase error across the transmitting antenna is the primary limitation of telescope performance, and optical telescopes are evaluated on the basis of their wavefront quality (the relative number of waves of phase error). The effect of wavefront error is both to flatten out the central lobe of the far-field pattern and to scatter energy out of the central lobe and into

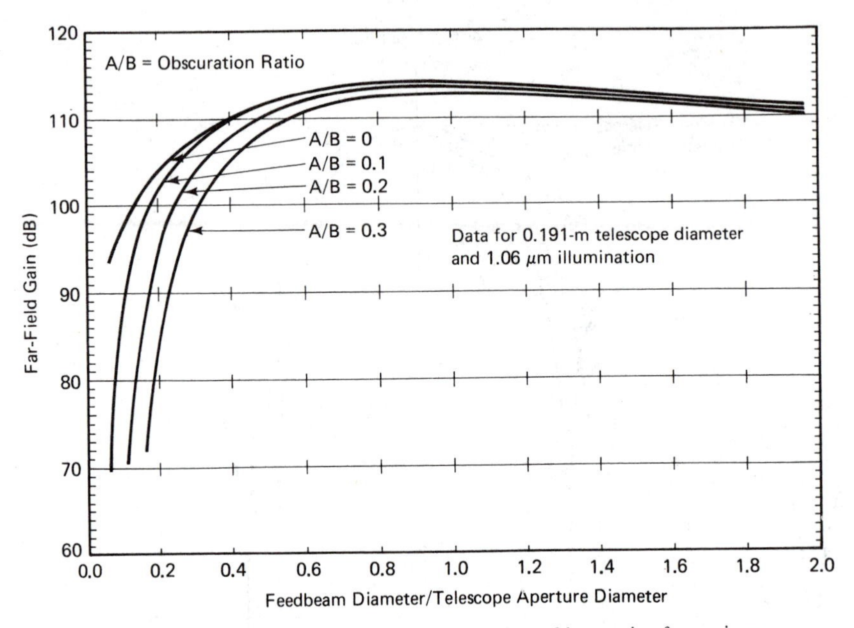

Figure 2-11 Curve shows antenna gain as function of beam size for various truncation to obscuration ratios (A/B).

the higher order sidelobes of the antenna pattern. This results in a reduction of the peak on-axis gain of the telescope antenna. The measure of efficiency, which is the ratio of on-axis antenna gain with wavefront error to that of an antenna with identical geometry without aberration often is referred to as the Strehl ratio.

Exact evaluation of Strehl ratio usually is limited to the most simple types of systematic types of aberration that can be expressed in a closed form mathematical expression. Generally, however, wavefront errors are considered to be randomly distributed over the antenna aperture. Under most conditions, the Strehl ratio is bounded by the relationship:

$$L_n = 10 \log_{10} e^{-(2\pi\eta/\lambda)^2} \tag{2-8}$$

where η = number of waves of rms phase error across the aperture ($\eta < 0.3\ \lambda$).

2.4.3 Receiver Antenna

The concept of receiver antenna gain is identical to that of the transmitter except in most laser communications systems the receiver acts primarily as an energy collector (incoherent receiver). In this case, the received optical energy incident on the receiving antenna is focused down on an optical detector usually much larger than the focused spot of light. Therefore, the receiver antenna generally is not sensitive to wavefront aberration. In addition, since it is typical for the receiver and transmitter to be separated by large distances (>5 km), the incoming radiation may be assumed to be a plane wave of flat phase and uniform amplitude distribution over the antenna surface. In this case, the receiver gain is given by:

$$G_r = 10 \log_{10}\left(\frac{2\pi a}{\lambda}\right)^2 \tag{2-9}$$

where a = antenna aperture radius.

2.4.4 Space Loss

Like RF systems, the concept of space loss refers to the fact that as the electromagnetic energy propagates in space the energy spreads out or dissipates with the phase front amplitude decaying. As with RF systems, the space loss is given by the expression:

$$L_R = 10 \log_{10}\left(\frac{\lambda}{4\pi R}\right)^2 \tag{2-10}$$

where R = link range between transmitter and receiver.

2.5 THE OPTICAL DETECTOR

In principle, the optical detector performs the same function in an optical communications system as its RF counterpart; i.e., it converts the incoming electromagnetic field to an electrical current that can subsequently be processed or demodulated by the receiving communications electronics. In general, however, because of the fundamental differences in the frequencies involved, optical detectors are typically square law devices—i.e., they produce an electrical current that is proportional to the incident energy of the field and not to the amplitude. Without going into a quantum mechanical explanation of optical detection, the general principle is most easily understood by recalling the corpuscular character of light where energy is contained in wave packets (photons) that have some momentum capable of exerting force on atomic electrons sometimes dislodging them to create photoelectrons. Consider, for example, the trivial optical detector of Figure 2-12 consisting only of a photosensitive metal surface (photocathode) and a collector (anode). Light striking the photocathode liberates photoelectrons from the cathode material, which are drawn to the anode due to their potential difference. The mean number of incident photons is given by:

$$N = P_{av} \times \frac{\lambda}{hc} \times T \tag{2-11}$$

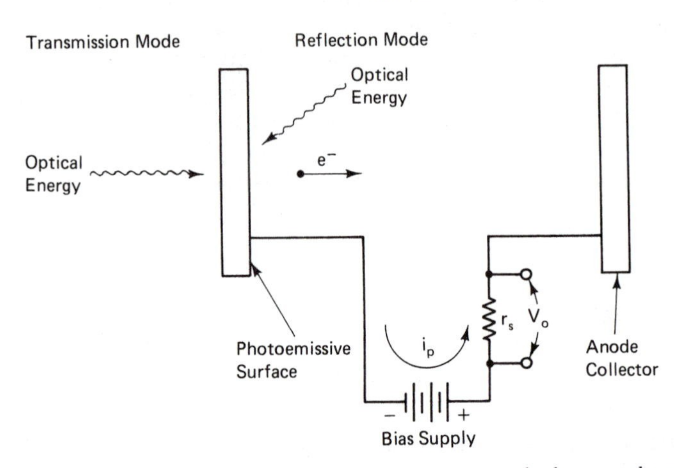

Figure 2-12 Optical photons impinging on photoemissive cathode cause the release of free electrons. These electrons are accelerated by the bias field to the anode. Current flowing through source resistor r_s results in output voltage proportional to input optical energy.

where

N = average number of photons
P_{av} = average incident optical power in watts
λ = free space wavelength of the optical energy
h = Planck's constant (6.6254×10^{-34})
c = speed of light in free space (2.997925×10^{8} m/s)
T = period of time of interest

The number of photoelectrons liberated at the photocathode as the result of the incident photons is dependent upon a number of physical characteristics of the material and window, which for our purposes are lumped into a constant known as the quantum efficiency of that device. Thus:

$$N_{pe} = N \times Qe \times q \tag{2-12}$$

where

N_{pe} = average number of photoelectrons
N = average number of photons
Qe = net quantum efficiency of the photocathode
q = electronic charge (1.602×10^{-19} Coulombs/electron)

One of the fundamental properties of optical communications is that transmission of optical energy is in itself a source of noise. Conceptually, this effect arises from a varying number of photons in each wave packet. This statistical process follows quantum mechanical principles and is characterized by conventional Poisson statistics. Thus, the probability of receiving a given number of photons or photoelectrons in any time interval is given by the equation:

$$P_N(n) = e^{-N}\left(\frac{N^n}{n!}\right) \tag{2-13}$$

where

$P_N(n)$ = probability of n photons in time interval T given an average of N photons of transmitted energy
N = average number of photons given by Equation (2-11)
n = number of photons in any time interval T

The fact that the signal energy itself is a statistical process is significant in that even in the complete absence of any channel background, thermal or detector noise there is maximum available signal to noise ratio useful for communications. It is easily shown that the signal to noise energy ratio is given by the equation:

$$SNR = \frac{N^2}{N} = N \tag{2-14}$$

where N = average number of received photons

For cases in which an optical background is present over the link, the signal to noise ratio is limited by the sum of the average signal and average background—i.e.,

$$SNR = \frac{Ns^2}{N_s + N_B} \tag{2-15}$$

where

N_s = average number of signal photons
N_B = average number of background photons

There are a variety of devices in use today as optical detectors. The most common types used for space optical communications are: the photomultiplier tube and two solid-state devices, the P-I-N, and the avalanche photodiode (APD). Each of these three are significant because each has a unique noise limited performance suited to particular applications.

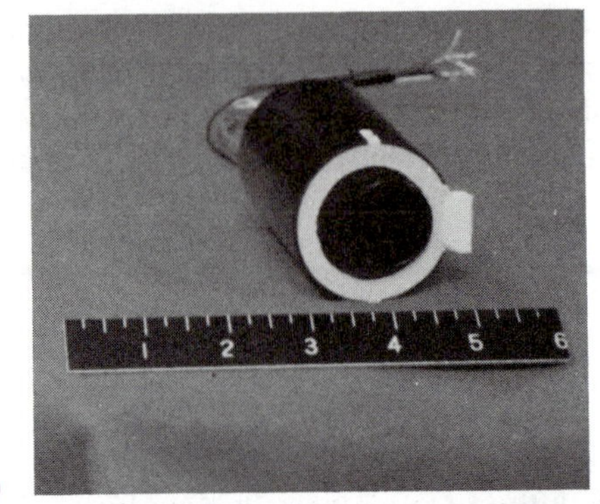

(a)

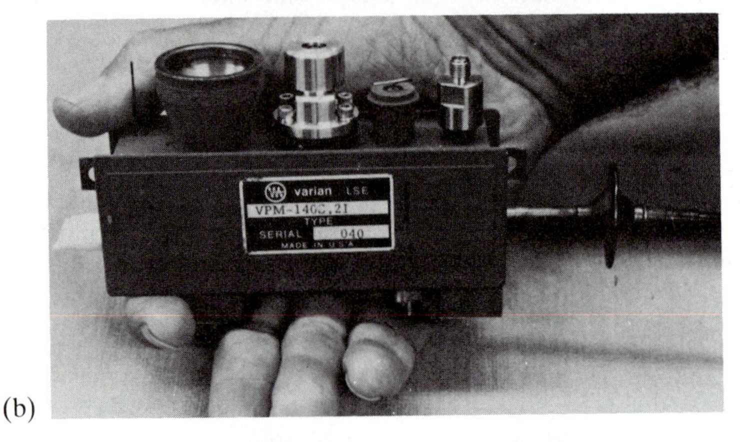

(b)

(c)

Figure 2-13 Various optical detectors: (a) Ruggedized photomultiplier tube—Glass envelope is replaced with metal and ceramic case. Internal elements are rigidly mounted to withstand space launch environments. (b) High-speed photomultiplier—Ruggedized tube designed for optical pulses <1 ns. External magnetic fields and internal r.f. electric fields work to produce ideal wideband "integrate and dump" operation. (c) Solid-state avalanche photodiode—with greater than 50-MHz bandwidth, this device including hybrid preamplifier is capable of optical performance superior to (a). Size and weight savings are obvious.

2.5.1 The Photomultiplier Tube

The various types of photomultiplier tubes shown in Figure 2-13 and represented schematically in Figure 2-14 are the closest to the ideal optical detector. In the photomultiplier tube, photoelectrons liberated from the photocathode are directed toward the first dynode. Each impact of a photoelectron with the dynode material results in the creation of several secondary electrons. The process is repeated for each of the several dynodes, until all electrons are finally collected by the anode. In this way, photomultipliers can typically achieve a gain of $>1 \times 10^6$. That is, for each photoelectron leaving the cathode as many as 1×10^6 photoelectrons are collected by the anode. To the first-order approximation, this gain is essentially noise-free, resulting in a statistical output identical to the input. Thus, signal-to-noise ratio is preserved through the tube. (There are a number of extraneous noise sources that do limit tube efficiency; they are covered in Chapter 4.) The effects of thermal noise in the follower resistor in the typical circuit application of Figure 2-14 are reduced by the gain of the photomultiplier. If the thermal noise variance of the resistor is given by i_{np}^2, it can be shown that the resulting signal-to-noise ratio for the photomultiplier is given to the first order by the expression:

$$SNR = \frac{N_s^2}{N_s + N_B + \frac{i_{np}^2}{G^2}\left(\frac{\tau}{q}\right)^2} \tag{2-16}$$

where

G = gain of the photomultiplier
τ = time interval ($\sim 1/BW$)
q = electronic charge

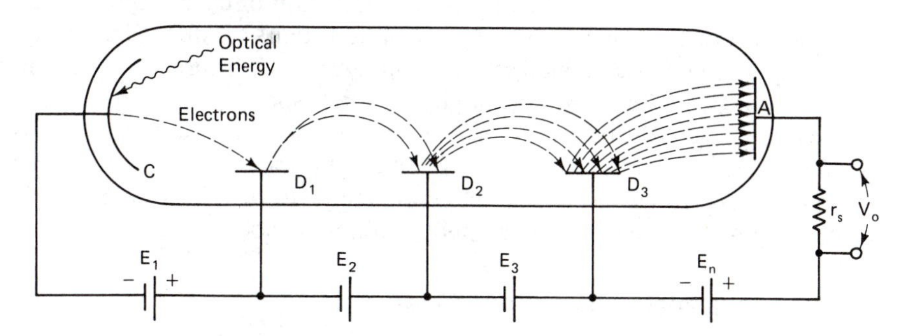

Figure 2-14 Schematic diagram of the photomultiplier tube showing electron emission multiplication. Photomultiplier tube usage is now usually restricted to applications where ultimate sensitivity over a large area is required. Typical electron gains range from 10^5 to 10^6 electrons/photoelectrons. Glass enclosure may be either evacuated or gas filled.

N_s = average signal photoelectrons
N_B = average background photoelectrons

The advantages of the noise-free PMT gain are obvious, making the follower resistor thermal noise negligible for moderate resistor values and typical PMT gains. However, aside from the physical limitations of size and the reliability of high-voltage components (5 kV or higher), photomultiplier tubes suffer from typically low quantum efficiencies. This is the result of limited materials and deposition techniques for the photoactive surfaces and the inability to collect efficiently all of the generated photoelectrons. In addition, each photoelectron follows a slightly different trajectory as it passes from photocathode to the dynodes and finally to the anode. Thus, there is a dispersion of arrival times for the various photoelectrons spawned off the photocathode. Usually the higher the gain, the greater the number of dynodes; the longer the pathlength, the greater the temporal dispersion. This dispersion, of course, lowers the usable communications bandwidth.

2.5.2 P-I-N Photodiodes

An alternative to the photomultiplier tube is the solid-state photodiode. These semiconductor devices are usually constructed on either silicon or AlGaAs substrates for visible or near infrared sensitive devices. These are two-port devices and electrically perform as a semiconductor junction, except when exposed to light. In operation, these devices, known as P-I-N (positive-intrinsic-negative) photodiodes because of their junction construction, are operated in a reverse-bias mode. Optical energy impinging upon the substrate material creates free electrons and holes. Again, the number of created electron-hole pairs is proportional to the number of incident photons. These free electron-hole pairs are swept into the junction by the field across the device. Upon entering the junction, they result in conduction of charge through the device proportional to the number of electron-hole pairs created. Electrically, the P-I-N photodiode is modeled as shown in Figure 2-15. There is no inherent gain in the device—i.e., one unit of charge is transported across the device for each generated electron-hole pair. Since the gain in Equation (2-16) is now unity, these devices are generally limited by the thermal noise in the electrical circuit around the device. However, these devices have significantly higher quantum efficiency than the photomultiplier tube.

2.5.3 Avalanche Photodiode (APD)

The avalanche photodiode (APD) is similar in construction to the P-I-N photodiode except that the doping of the junction material is altered, which changes the voltage versus current slope characteristic in the vicinity of the reverse-bias avalanche breakdown. When this device is biased near the

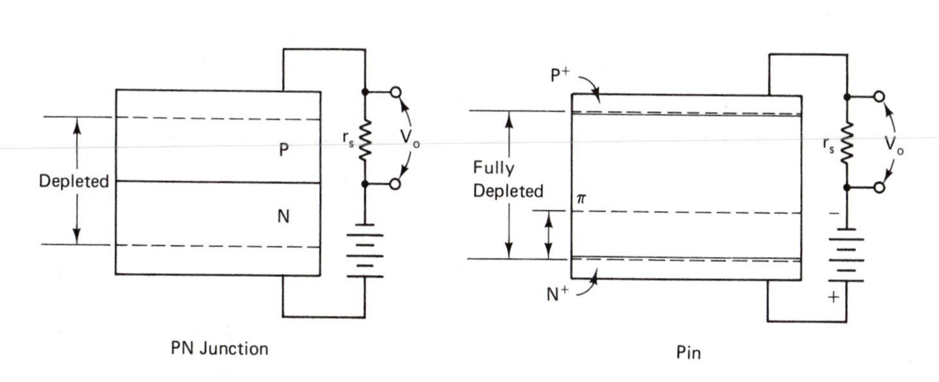

Figure 2-15 Two Types of Solid State Junction Photodiodes. In both devices photons intersect with bulk material to liberate electrons and holes. These carriers migrate into the field region where they are swept across the junction resulting in current flow. The APD is a special type of P-I-N photodiode.

breakdown voltage, optically generated electron-hole pairs entering the junction stimulate the generation of many additional carriers giving the effect of avalanche gain to the device. Typical values for avalanche gain usually are less than 200, and unlike the PMT this gain cannot be assumed to be noise-free.

Analytically, APD multiplication can be described as the probability of getting a pulse of m electrons out of the device when exactly n photoelectrons are injected into the gain region of the diode. Mathematically, this probability is predicted by the rather complicated expression.[13]

$$P_{n,m} = \frac{n\Gamma(m/1-k+1)}{\mathrm{m(m - n)!}\Gamma\left(\dfrac{km}{1-k}+1+n\right)} \times \frac{1+k(M-1)^{(n+km/1-k)}}{M} \times \frac{(1-k)(M-1)^{(m-n)}}{M} \tag{2-17}$$

where $\Gamma(\;)$ is the gamma function, k is the ratio of ionization coefficients for holes and electrons, and M is the average gain. Figure 2-16 is a plot of Equation (2-17) for various values of incident photoelectrons, n, showing the unique probability distribution of multiplication gain.

Even though the APD has no photocathode per se, there is a region where incident optical power is converted to electrical carriers via the generation of free electrons and holes. Just as with the actual photocathode of a PMT, this carrier generator is a Poisson shot-noise process. That is, the probability of getting a pulse of n photoelectrons from the photoemissive surface is given by Equation (2-13).

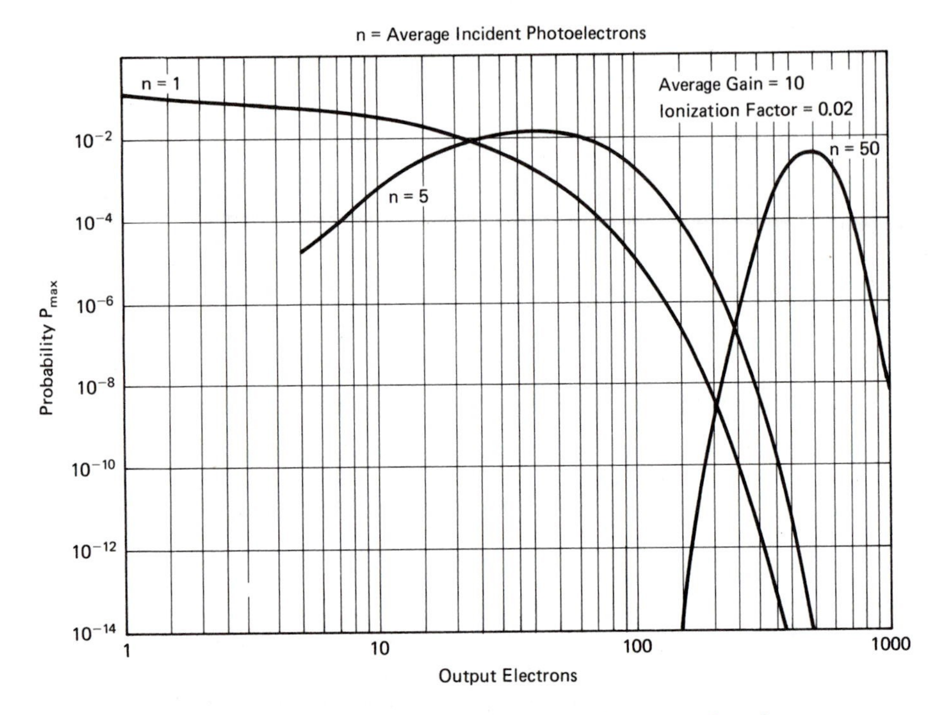

Figure 2-16 APD Multiplication Gain Distribution. The curves show the combined effects of shot noise and APD gain probabilities for various mean rates of incident photoelectrons. Note that for a mean incident rate of 50 the output distribution is nearly gaussian with mean centered at 500 (average incident mean rate X average gain). In this case the variance is predicted very closely by Equation 2-19.

The probability of obtaining m electrons out of an APD with a mean rate of N incident photoelectrons in a pulse is given by

$$P_N(m) \sum_{n=1}^{\infty} P_N(n)\, P_{n,m} \tag{2-18}$$

where $P_N(n)$ is given in Equation (2-13), and $P_{n,m}$ is from Equation (2-17).

Equation (2–18) is cumbersome and unfortunately no approximations have been found for small mean rates of incident photoelectrons. Thus, the probability must be evaluated numerically for conditions of low gains and low incident rates of photoelectrons.

For high mean rates, however, Equation (2-13) becomes very nearly Gaussian with a variance of N. Additionally, for large values of n, $P_{n,m}$ falls off very rapidly for m much different from nM, making it essentially a Gaussian distribution over the range of probabilities of interest. Combining both of those expressions yields the simplifying approximation.

$$P_N(m) = \frac{1}{(2\pi\sigma^2)^{1/2}}\, e^{-(m-NM)^2/2\sigma^2} \quad N/F >> 1 \tag{2-19}$$

where $\sigma^2 = NM^2F$ and F is the excess noise factor given by

$$F = kM + (1 - k)(2 - 1/M) \tag{2-20}$$

Dark noise

In addition to multiplication noise and quantum noise, the third source of noise from a detector is dark current. Dark noise arises from two sources: surface leakage currents, which do not undergo APD multiplication; and bulk leakage current, which is subject to gain multiplication. The noise arising from both of those sources is a shot noise and is evaluated by the same techniques as quantum noise.

Preamplifier noise and optimum gain

The fourth noise source encountered using the APD is that of the associated preamplifier. For convenience of analysis, the preamplifier noise can be related to an effective mean rate at the photocathode by the expression

$$N_P = \frac{i_p^2}{M^2}\left(\frac{\tau}{q}\right)^2 \tag{2-21}$$

where

i_p^2 = variance of the transimpedance preamplifier referenced to a current input
τ = laser pulsewidth in seconds, assuming matched filter detection
q = electronic charge
M = APD gain
N_P = effective mean rate at the photocathode in photoelectrons/pulse

From Equation (2-21), it can be seen that the effect of preamplifier noise on system performance varies as the inverse square of the detector gain just as with the PMT. As would be expected, the higher the gain, the less the effect of preamplifier noise.

However, making the effect of preamp noise arbitrarily small by increasing avalanche gain usually is not practical, since multiplication noise (which is a direct function of gain) becomes the limiting parameter.

In the absence of optical input, the total noise referenced to the effective photocathode is given by

$$N_T = \frac{I_{DS}\left(\frac{\tau}{q}\right)}{M^2} + I_{DB}\left(\frac{\tau}{q}\right)F + \frac{i_p^2}{M^2}\left(\frac{\tau}{q}\right)^2 \tag{2-22}$$

where

I_{DS} = dark surface current
I_{DB} = dark bulk current
M = average APD gain
F = APD excess noise factor

Figure 2-17 is a plot of N_T versus M for typical values of detector dark noise and preamplifier noise. From Figure 2-17, it can be seen that the optimum APD gain is a function of the anticipated optical background as well as detector dark noise and preamplifier noise. If optimum performance of the detector is desired under all conditions, some form of gain control must be employed to adjust the APD gain as a function of background. In the limit, the avalanche gain goes to unity for high optical backgrounds and the device functions as a P-I-N device.

2.5.4 Receiver Design

Once the detector has been selected for the optical communications system, the electrical output from the detector must now be processed to extract the transmitted information out of all of the various noise sources. The conventional method to accomplish this is with a simple threshold discriminator circuit (Figure 2-18). This detection process often is referred to as direct detection. At the discriminator, a threshold is set so that the probability of a false alarm P_{fa} (false detection) is reduced to an acceptable level while minimizing the required signal, N_s, to establish the desired detection probability, P_d. The relationship between P_{fa} and P_d is seen in Figure 2-19.

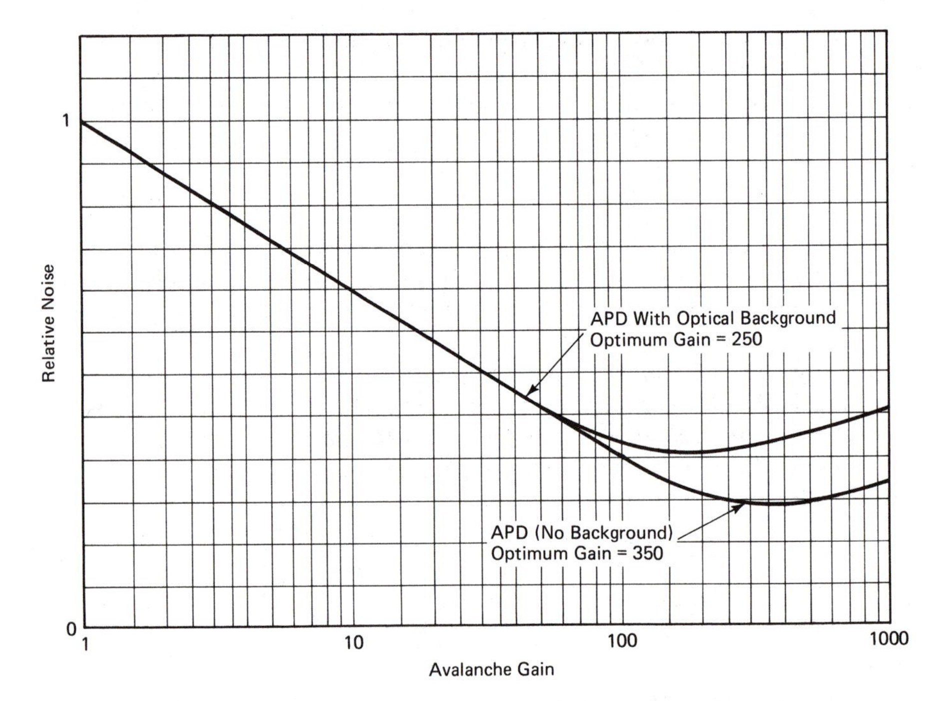

Figure 2-17 Plots of Equation (2-22) with and without optical background. Curves illustrate that even though APD multiplication gain is noisy, significant performance improvement can be realized over P-I-N photodiodes that operate at unity gain.

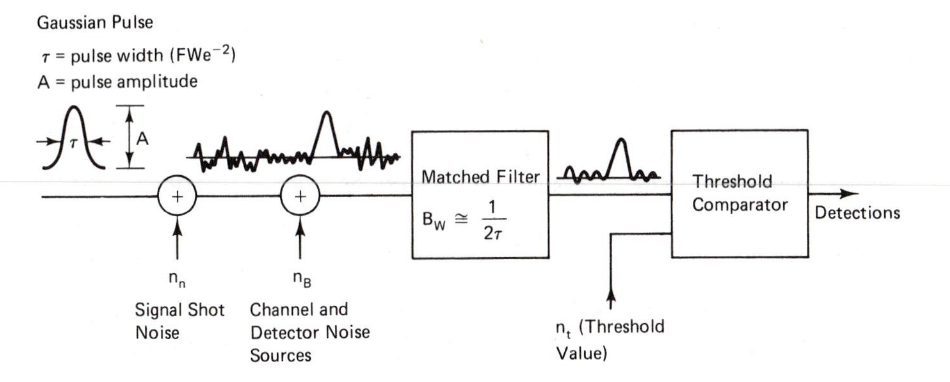

Figure 2-18 Standard model of communication link and receiver. Most noise sources are additive except APD multiplication gain. The matched filter assumes a fixed correlation internal equal to pulsewidth. A detection flag occurs whenever input signal or noise exceeds the noise threshold of the comparitor.

Threshold determination

A detection threshold is set on the basis of noise from three sources: (1) optical background, (2) detector dark noise, and (3) preamplifier noise. (APD multiplication noise contributes to the background noise and detector noise.) An exact analysis of the noise threshold setting is difficult, since the probability distributions associated with these noise sources is often different.

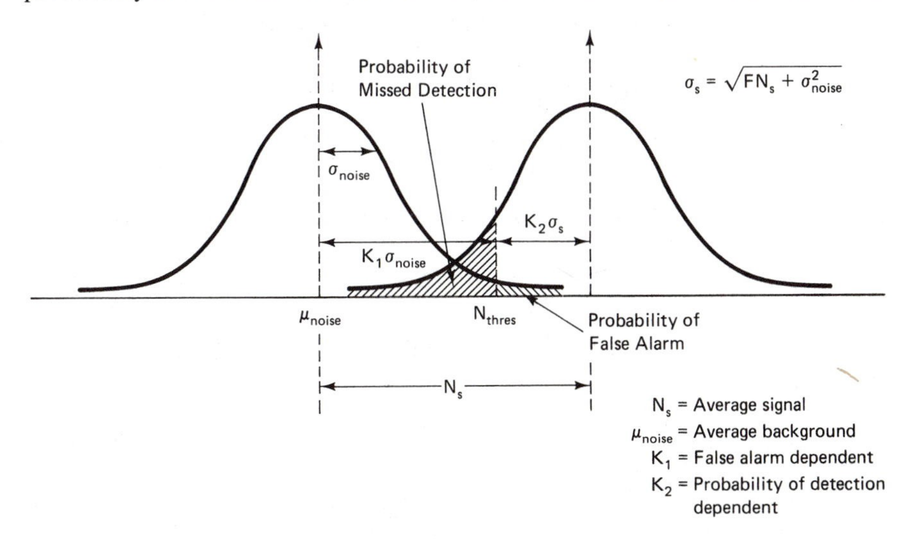

Figure 2-19 The noise and signal probability density curves illustrate the detection process. The probability of false alarm, P_{fa}, is the area of the noise density curve *above* the threshold. The probability of missed detection, $I - P_d$, is the area of the signal + noise density curve below the threshold. In optical communications, the noise variance with signal is greater than the noise variance without signal.

Background and dark noise distributions are predicted by Equation (2-18), and the preamplifier noise is typically whole Gaussian distributed. For most detectors (including APDs) operating at moderate gains, all noise sources may be considered Gaussian with a probability distribution given by Equation (2-19). Making this assumption facilitates evaluation of the total noise equivalent power NEP. [The Gaussian approximation is a conservative approximation in that the total Gaussian noise variance given by $\tau^2 = NM^2F$ is greater than the actual APD multiplication variance predicted by Equation (2-18).]

Using Gaussian assumptions, the effective noise power (in the absence of signal) referenced back to the hypothetical photocathode becomes:

$$N_{TV} = \frac{I_{DS}}{M^2}\left(\frac{\tau}{q}\right) + F(I_{DB} + I_B)\frac{\tau}{q} + \frac{i^2_{na}}{M^2}\left(\frac{\tau}{q}\right)^2 \tag{2-23}$$

where

I_{DS} = APD dark surface current
I_{DB} = APD dark bulk current
I_B = background current
B_W = noise bandwidth
M = average APD gain
F = APD excess noise factor
i^2_{na} = equivalent preamplifier noise input power
N_{TV} = average noise rate (in photoelectrons/s)

The mean noise rate at the photocathode is given by

$$N_{Tm} = \frac{I_{D_S}}{M^2}\frac{q}{\tau} + F(I_{D_B} + I_B)\frac{q}{\tau} \tag{2-24}$$

From Equations (2-23) and (2-24), a detection threshold is set so that

$$1 - \int_{\infty}^{N_{TH}} \frac{1}{2\,N_{TV}}\, e^{-(m-N_{Tm})^2/2N_{TV}}\, dm \leq P_{fa} \tag{2-25}$$

where N_{TH} is the detection threshold and P_{fa} is the desired probability of false alarm.

Required signal

Once a threshold has been determined on the basis of noise, the required signal can be determined, making the same Gaussian assumption as in the previous case. The signal noise variance is given by

$$\sigma_s^2 = F(I_{\text{sig}} + I_B + I_{DB})\frac{q}{\tau} + \frac{i^2_{na}}{M^2}\left(\frac{q}{\tau}\right)^2 + \frac{I_{DS}}{M^2}\left(\frac{q}{\tau}\right) \tag{2-26}$$

where I_{sig} is the mean of the required signal.

The mean is given by

$$\mu = F(I_{\text{sig}} + I_B + I_{DB})\frac{q}{\tau} + \frac{I_{DS}}{M^2}\left(\frac{q}{\tau}\right)$$

The incident signal must be sufficient to satisfy the relation

$$1 - \int_{\infty}^{N_{TH}} \frac{1}{2\pi\sigma_s^2} e^{-(m-\mu)^2/2\sigma_s^2}\, dm \geq P_D \tag{2-27}$$

where P_D is the required probability of detection.

2.6 OPTICAL MODULATION FORMATS

The analysis of RF modulation techniques properly assumed thermal noise as the primary system noise source. At optical frequencies, as just seen, this is not the case and the conclusions regarding system modulation effectiveness for RF frequencies are not necessarily valid at laser frequencies. In fact, consideration of the quantum effect, in conjunction with optical detection techniques, leads one to derive modulation techniques for laser communications that would be of questionable advantage at RF.

For most high data-rate laser communication formats the short pulses with high peak powers and low duty cycles inherent in pulsed lasers are particularly appealing. Three short pulse laser modulation formats that are shown as examples typical to high data-rate communication systems are:

- Pulse-Gated Binary Modulation (PGBM)
- Pulse Polarization Binary Modulation (PPBM)
- Pulse Interval Modulation (PIM)

2.6.1 Pulse-Gated Binary Modulation

In the Pulse Gated Binary Modulation (PGBM) format, a train of periodically occurring pulses are gated on or off so that a binary "1" is represented by the presence of a laser pulse, and a binary "0" is represented by its absence. Figure 2-20 shows a PGBM laser communication system basic block diagram and a representative waveform. PGBM is similar to normal CW binary PCM, except that PGBM is used in a laser system that operates at a low duty cycle (i.e., ratio of pulse width to pulse spacing is $<< 1$) because of the inherently short pulse width of pulsed laser operation. The PGBM format is compatible with the inherently periodic pulse output of a modulated laser. In addition, fixed pulse spacing of the PGBM laser system permits the use of a pulse-gated or "integrate and dump" optical receiver for noise discrimination. However, because transmitter energy is suppressed for roughly

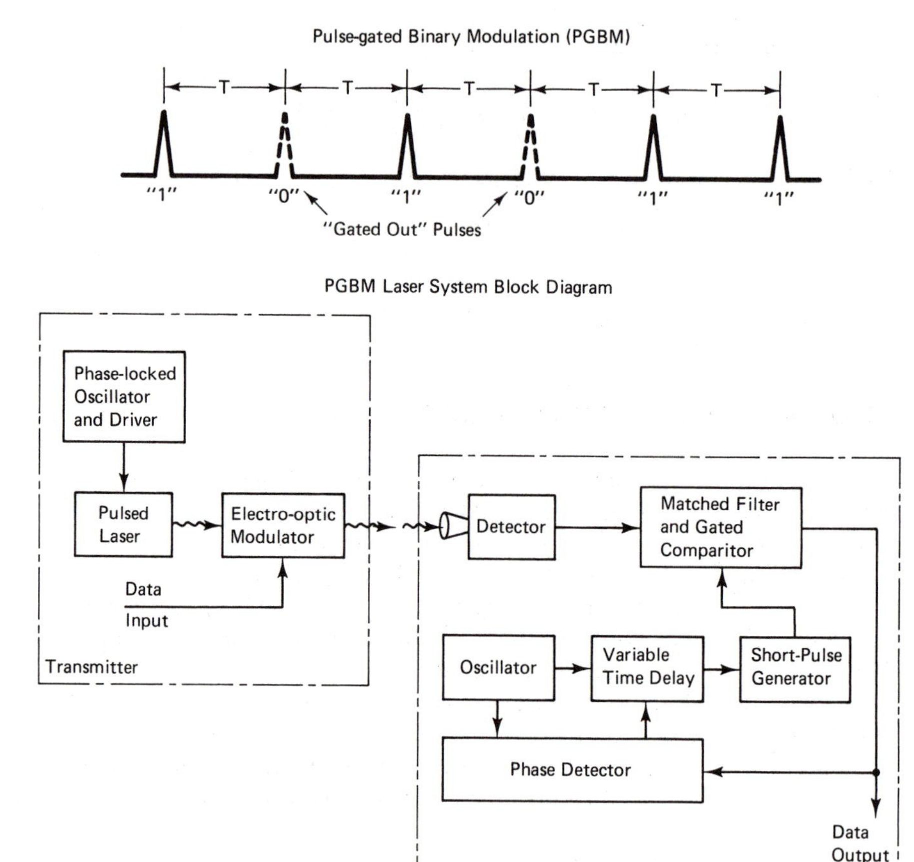

Figure 2-20 Pulse-gated binary modulation (PGBM): Information is conveyed by the presence or absence of an optical signal in bit period T. Since no signal is transmitted for a "0," the signal to noise ratio for 1 and 0 are not symmetric. In addition, half of the available transmitter is suppressed (for 0's) making the format very inefficient.

1/2 of the bits transmitted (the 0's), the format is inefficient from the standpoint of transmitter power utilization.

2.6.2 Pulse Polarization Binary Modulation

For Pulse Polarization Binary Modulation (PPBM), instead of gating the laser pulses on and off (as in PGBM), the polarization of the laser pulse radiation is rotated 90° for differentiation of the 1 and 0. An advantage of PPBM is that a pulse is expected during every bit interval at the receiver improving both noise performance and transmitter efficiency. A disadvantage is that the receiver must have dual channels, one for each polarization.

2.6.3 Pulse Interval Modulation

Pulse Interval Modulation (PIM) is an M-ary format in which one pulse conveys information representing many bits of ordinary binary code. The normal interpulse time interval is divided into M discrete time slots, and one and only one pulse is sent in the time interval of M slots. Figure 2-21 shows a PIM laser communication system basic block diagram and a representative waveform.

To encode information in the PIM format, binary data is grouped into words with word length equal to the transmitted bits/pulse. The value of the word is then converted to a unique time slot for pulse transmission. In this way, each pulse represents a word comprised of several bits. The number of bits transmitted per pulse is $\log_2 M$, where M is the number of discrete time slots per interpulse intervals.

In a short pulse laser system, many time slots are possible in a given time interval, permitting many bits per pulse. Formats transmitting up to 7 bits/pulse (128 discrete slots) are typical depending on data rate. In addition, very low duty cycle operation is achieved since only one pulse occurs during the M time intervals. This duty cycle is approximately $1/M$, but the exact value is dependent on whether or not the laser pulse width is less than the slot width and how often a synchronization timing pulse must be used.

The PIM format permits temporal noise discrimination by using discrete sampling of each narrow time slot in the optical receiver. This capability, along with the advantages of many bits per pulse, make the PIM laser system potentially the most efficient. One disadvantage of PIM is that an incorrect detection of a single pulse can result in the loss of several consecutive bits. For this reason, PIM systems usually employ convolutional coding techniques so that word errors are distributed randomly in the data stream.

2.7 DERIVING ERROR RATE STATISTICS

Evaluation of communications link performance is usually done by establishing a bit error rate or a probability of bit error for message transmission. As discussed, all communications links are limited by noise. For RF systems, the limiting noise contributors are ambient background radiation and thermal noise in the receiving electronics. For optical systems, the noise sources include optical background radiation, signal quantum noise, receiver thermal noise, and detector characteristics.

The function of the receiver is to correctly detect the message in the presence of this noise. For PGBM this means correct detection of the presence

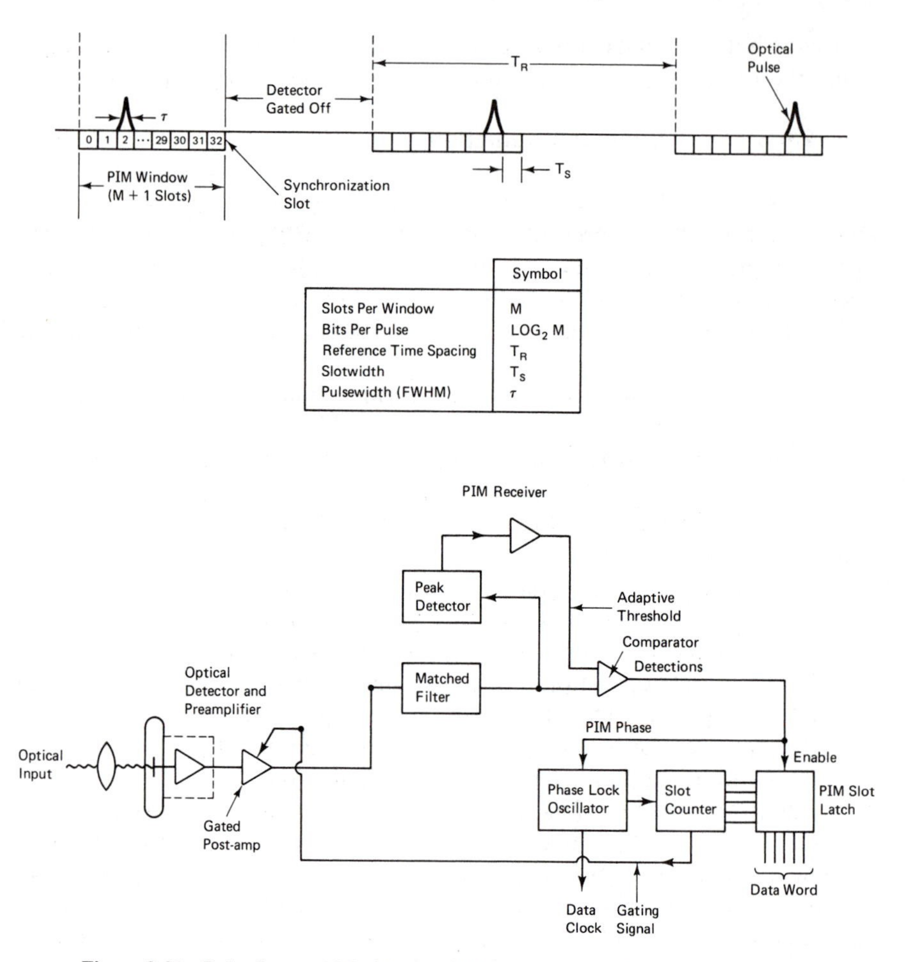

	Symbol
Slots Per Window	M
Bits Per Pulse	LOG_2 M
Reference Time Spacing	T_R
Slotwidth	T_S
Pulsewidth (FWHM)	τ

Figure 2-21 Pulse Interval Modulation (PIM) Format and Receiver: Two options exist for PIM. Data can be transmitted by modulating the time interval between two pulses or by modulating the transmission time of a single pulse with respect to a fixed reference. Although the latter offers improved performance (BER improves by $2X$ for fixed SNR), the cost is increased receiver complexity. In both cases, performance is usually diminished by P_{fa}. The optimum threshold is set to minimize P_{fa}. Since a false alarm or missed detection results in missing M bits at a time, this m-ary format is subject to "burst errors." Detector gating is included to reduce false alarms between PIM windows.

of a signal pulse during a bit period indicates a "1" transmitted and not detecting a pulse during the bit period is detected as a "0." Referring back to Equation (2-19), the probability of receiving m photoelectrons given a "1" transmitted is given by equation:

$$P(m/1) \sim \frac{1}{2\pi\sigma_s^2}\, e^{-(m-N_sM)^2/2\sigma_s^2} \tag{2-28a}$$

where

σ_s^2 = noise variance with signal and background at the receiver [Equation (2-26)]
N_s = mean number of received signal and noise photoelectrons
M = average detector gain

Similarly the probability of receiving m photoelectrons given a "0" transmitted is given by the expression:

$$P(m/0) \sim \frac{1}{2\pi\sigma_B^2} e^{-(m - N_B M)^2/2\sigma^2} \tag{2-28b}$$

where

σ_B^2 = noise variance with background only at the receiver [Equation (2-23)]
N_B = mean number of background photoelectrons [Equation (2-24)]
M = average detector gain

Let $P(0)$ be the *a priori* probability of transmitting a "0" and $P(1)$ is the *a priori* probability of transmitting a "1." If $P(0) = P(1)$, the receiver will decide a "1" has been sent if $P(m|1) \geq P(m|0)$. If $P(0) \neq P(1)$, the detector will decide a "1" has been sent if $P(1)P(m|1) \geqslant P(0)P(m|0)$. This last relationship constitutes what is known as the "ideal observer criterion."

For some value of m, say $m = n_t$, the ideal observer criterion equality will be satisfied, and it can be written as

$$\frac{P(n_t|1)}{P(n_t|0)} = \frac{P(0)}{P(1)} \tag{2-29}$$

The value of $m = n_t$ is a threshold value because each time the receiver photodetector counts less than n_t photoelectrons it will decide a "0" was sent, and each time greater than n_t photoelectrons are counted by the receiver it will decide a "1" was sent.

In order to derive the optimum threshold, n_t, Equation (2-28a) and Equation (2-28b) are substituted into Equation (2-29) for the special case $m = n_t$. Equation (2-30) results from making these substitutions and rearranging the expression slightly.

$$\frac{(N_s + N_B)^{n_t}}{(N_B)^{n_t}} e^{-N_s} = \frac{P(0)}{P(1)} \tag{2-30}$$

Solving Equation (2-30) for n_t results in the expression in Equation (2-31).

$$n_t = \frac{N_s + ln\left(\frac{P(0)}{P(1)}\right)}{ln\left(\frac{N_s + N_B}{N_B}\right)} \tag{2-31}$$

In actuality, for most binary coded information, the average number of 1's and 0's will be the same. Hence $P(0) = P(1) = 1/2$, and the derivation of Equation (2-31) is applicable to threshold detection in general, rather than PGBM in particular.

In general, n_t may not be an integer, so define N_T as the greatest integer less than or equal to n_t. Then the minimum probability of error in signal detection can now be found. P_{FD} is the probability of false detection and can be described as the probability of detecting a "0" as a "1." P_{ND} is the probability of a missed detection and is described as the probability of detecting a "1" as a "0." The probability of error, P_E, is:

$$P_E = P_{ND}P(1) + P_{FD}P(0) \tag{2-32}$$

When a "1" is sent, the probability of counting m photoelectrons is given by Equation (2-28a). However, each time $m < N_t$, the photodetector will decide that a "0" has been sent. Hence,

$$P_N = \int_{-\infty}^{N_t} \frac{1}{2\pi\sigma_s^2} e^{-(m - N_s M)^2/2\sigma_s^2}\, dm \tag{2-33}$$

Likewise, when a "0" is sent, the probability of counting m photoelectrons is given by Equation (2-28b). However, each time $m \geq N_t$, the photodetector will decide a "1" has been sent. Hence,

$$P_{FD} = 1 - \int_{-\infty}^{N_t} \frac{1}{2\pi\sigma_B^2} e^{-(m - N_B M)/2\sigma_B^2}\, dm \tag{2-34}$$

Substituting Equations (2-33) and (2-34) into Equation (2-32) yields the expression for probability of error.

$$\begin{aligned} P_E = {} & P(0) \int_{-\infty}^{N_t} \frac{1}{2\pi\sigma_B^2} e^{-(m - N_B M)/2\sigma_B^2}\, dm \\ & + P(1) \int_{-\infty}^{N_t} \frac{1}{2\pi\sigma_B^2} e^{-(m - N_B M)/2\sigma_B^2}\, dm \end{aligned} \tag{2-35}$$

2.7.1 Derivation of Probability of Error For PIM

In the PIM format, information is transmitted by marking the occurrence of the pulse in one of several discrete time slots within the interpulse time interval. The detector, therefore, must decide in which slot the pulse was transmitted, and this can be done in one of two ways. The detector can compare the photoelectron count in each slot either with a preset threshold or with the counts in all the other slots.

In the PIM case, in which a threshold detector is utilized, it is assumed that in the interpulse time interval, T_p, there are M equal periods of duration G. Since there is one and only one signal pulse of width, T, transmitted in each time interval T_p, the probability that the pulse occurs in a given slot is

$1/M$. Thus, per interval G, $P(1) = 1/M$. The probability that the pulse does not occur in a given slot is then $1 - 1/M$ because $P(1) + P(0) = 1$ always. Hence, $P(0) = 1 - 1/M$ per interval G. The optimum threshold derived for the PGBM case may be applied to each interval G, in the PIM case, by using the above values of $P(1)$ and $P(0)$, and Equation (2-31).

$$N_t = \frac{N_s + \ln (M)}{\ln \left(\dfrac{N_s + N_B}{N_B} \right)} \tag{2-36}$$

P_{FD} and P_{ND} per interval G in this PIM threshold case are of the same form as derived for the PGBM case. However, the formulation of PIM threshold probability of error is somewhat different. In order to detect a PIM threshold pulse correctly, one slot of width G must exceed the threshold and $(M - 1)$ slots must not exceed the threshold. The probability of correct detection is the product of these M decisions being made correctly. The probability of error is, therefore

$$P_E = 1 - (1 - P_{ND})(1 - P_{FD})^{(M-1)} \tag{2-37}$$

2.7.2 Derivation of Probability of Error For PPBM

In PPBM modulation, the information pulse occurs in one of two polarization states, so the receiver must compare the relative energy of the two detectors and select the greatest. This is often referred to as the "greatest of" technique. Mathematically, PPBM is a special case of PIM using $M = 2$ with the substitution of $M = 2$ into Equation (2-37) a comparison between PIM and PPBM can be made as shown in Figure 2-22.

2.8 SIGNAL REQUIREMENTS FOR ACQUISITION AND TRACKING[14]

Signal requirements for acquisition and tracking are driven both by required system detection statistics (false alarm rate (FAR) and probability of detection (P_D)) and by required minimum tracking error. For the laser crosslink acquisition sequences, the dominant requirement usually is signal detectability. However, for tracking, angle noise is the driving signal requirement.

The number of photoelectrons per pulse required during the acquisition sequence is determined by detector false alarm rate (FAR) and detection probability (P_D). Typical system requirements for these parameters are a FAR of 1/s, and $P_D = 90$ percent per pulse so that for an acquisition detector there will be on the average one false alarm every second and 1 missed detection in a second. The detection logic consists of an ungated threshold comparator, with the threshold set for a one per second FAR.

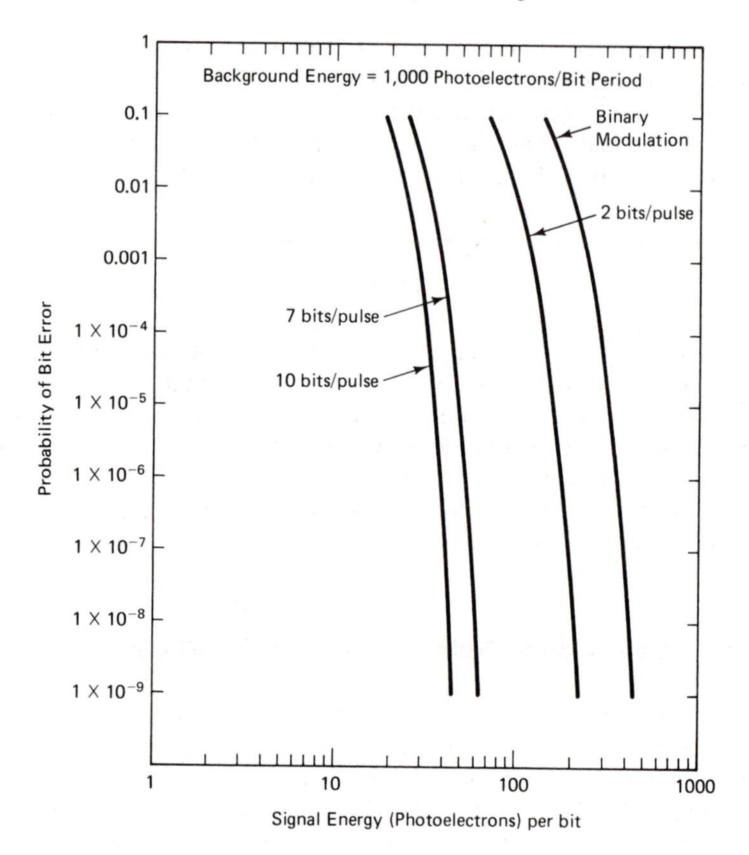

Figure 2-22 Bit error rate performance is shown as a function of received optical energy per bit. The advantage of high order *m*-ary modulation formats over binary formats is shown by the lower probability of bit error for a constant signal for the higher order formats. The system designer must weigh trade-offs between data rate, system bandwidth, and laser output power to determine the correct format.

Although the input signal to the threshold circuit is ungated, it is convenient to assume a block correlated input that consists of n statistically independent samples of width, τ, where τ is the correlation time usually matched to the received pulse width by the relationship:

$$BW \simeq 0.4/\tau \tag{2-38}$$

The number of independent opportunities, then, for noise pulses to cross the threshold in a second (false alarms) is less than $BW/0.4$. The probability of false alarm P_{fa} is the probability of a false alarm per opportunity and given by the expression:

$$P_{fa} = \frac{0.4\ FAR}{BW} = P(m > TH) \tag{2-39}$$

where m is the received background and noise photoelectrons in τ, and TH is the threshold of the comparator where P_{fa} is defined by Equation (2-25), P_D is the probability of detecting the transmitted pulse and defined by Equation (2-27).

During the tracking mode the required signal is set by the need for a high signal-to-noise ratio on the processed angle output rather than by detection probabilities. The angle noise from the processor is given by the ratio of the noise voltage at the processor output divided by the scale factor of the angle-measuring process in volts per unit angle. Since for most applications the system is aligned near boresight, the angle noise is computed at boresight.

The quadrant detector as illustrated in Figure 2-23 is used for vertical position sensing by subtracting the energy found in the lower two quadrants from that in the upper two and dividing by the sum of all four.

Similarly, the horizontal spot position is found by subtracting the energy in the left quadrants from that in the right and again dividing. The division process normalizes the measurement with respect to total received signal, effecting a sample-by-sample automatic gain control.

Notice that in Figure 2-24 a region can be drawn such that the illumination within that region corresponds to the difference between the upper and lower detector halves. Consequently, the ratio of this difference to the total illumination is given by the equation:

$$f(Y) = \left. \frac{\int_{-Y}^{Y}\int_{-\infty}^{\infty} I(x,y)dxdy}{\int_{-\infty}^{\infty}\int_{-\infty}^{\infty} I(x,y)dxdy} \right|_{y=Y} \tag{2-40}$$

where $I(x,y)$ is the power distribution over the detector in watts per unit area, and y is the displacement of the spot center from the horizontal dividing line

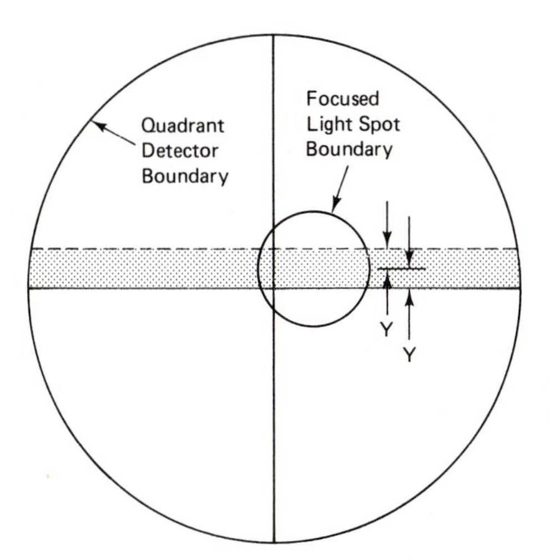

Figure 2-23 Quadrant detector geometry: The focused spot center is at location (x,y) relative to the quadrant axes. The shaded region corresponds to area difference between upper and lower detector halves. © IEEE.

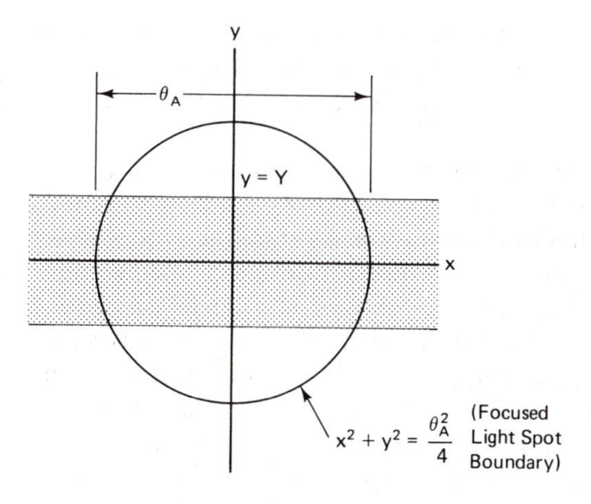

Figure 2-24 Geometry for computing $I(y)$. Translated coordinates with origin at spot center facilitate evaluation at shaded region of Figure 2-23. ©IEEE.

between the quadrants. A similar expression represents the lateral error function. The infinite limits are used for convenience under the assumptions that the spot is much smaller than the detector so that $I(x,y)$ is essentially zero at the detector boundary.

The sensitivity at $Y = 0$ is directly related to the angular scale factor of the detector. The sensitivity is given by

$$\left.\frac{df}{dY}\right|_{Y=0} = \left.\frac{\dfrac{d}{dY}\displaystyle\int_{-Y}^{Y}\int_{-\infty}^{\infty} I(x,y)dxdy}{\displaystyle\int_{-\infty}^{\infty}\int_{-\infty}^{\infty} I(x,y)dxdy}\right|_{y=Y} = \frac{2\displaystyle\int_{-\infty}^{\infty} I(x,0)}{P} \tag{2-41}$$

where P is the total energy in the spot.

For a uniform circular spot of intensity I_0 and i/e^2 diameter d_1, the numerator is given by

$$2\int_{-\infty}^{\infty} I(x,0) = 2I_0 d_1$$

and the total power by

$$P = \frac{\pi d_1^2}{4} I_0$$

Therefore,

$$\left.\frac{df}{dy}\right|_{y=0} = \frac{8}{\pi d_1} = \frac{2.55}{d_1}$$

For an Airy distribution (diffraction-limited optics) of peak intensity I_0 and Airy disk diameter D_A, the numerator is given by:

$$2\int_{-\infty}^{\infty} I_0 \left[\frac{2f_1(h)}{h}\right]^2 dx$$

where

$$h = \frac{2.44\pi x}{D_A}$$

The denominator is given by

$$P = \int_0^{\infty} I_0 \left[\frac{2f_1(g)}{g}\right] 2\pi r dr$$

where

$$r^2 = x^2 + y^2$$

$$g = \frac{2.44\pi r}{D_A}$$

By mathematical manipulation, the ratio reduces to

$$\left.\frac{df}{dy}\right|_{y=0} = \frac{(16)(2.44)}{3\pi D_A} = \frac{4.14}{D_A}$$

To compute the angle noise, the sensitivity computed above must be divided into the noise in the estimate of the ratio. The Quadrant detector has four channels processed in a difference over sum manner. The noise in each channel is produced by the preamplifier for each channel i_p referenced to the optically produced signal current. The numerator in the ratiometric processor has a value at boresight whose deterministic component is zero, and whose random component is the sum of four independent Gaussian sources. This amounts to

$$i_n^2 = i_{n1}^2 + i_{n2}^2 + i_{n3}^2 + i_{n4}^2 = 4i_p^2$$

For a reasonable signal-to-noise ratio, the noise component in the denominator can be ignored and the variance in the ratio of Equation (2-40) computed as

$$\sigma_r^2 = \frac{4i_p^2}{I_s^2} = \frac{4}{SNR}$$

where *SNR* is defined as total signal power over single-channel noise power in the bandwidth of the preamplifier.

The standard deviation on the position error for this Airy disk case is then given by

$$\frac{\sigma_r}{df/dy} = \frac{2}{\sqrt{SNR}} \frac{3\pi d_A}{(1.6)(2.44)} = 0.483 \frac{d_A}{\sqrt{SNR}}$$

This random position error is converted into an angle error by dividing by the focal length f of the optical system. Given

$$\theta_A = \frac{d_A}{f}$$

then

$$\sigma_\theta = 0.483 \frac{\theta_A}{\sqrt{SNR}}$$

is the angle noise of the system. In addition, all optical systems suffer from some coalignment error between transmit and receive paths due to finite capabilities for measuring and aligning optical systems. Thus, the total pointing error may be assumed to be bounded by the equation

$$\theta_\varepsilon = 3\tau_\theta + \theta_B \tag{2-42}$$

where θ_B is the static alignment bias of the optical system.

2.9 FUNDAMENTALS OF SYSTEM DESIGN

Having reviewed the key components and concepts of a space laser communications link, the task befalls the system designer to judiciously allocate system performance to the various components. As with any spaceborne communications package, the most obvious design trades involve the tangible parameters: performance, size, weight, and power consumption. However, because laser communications systems today utilize many new technologies (to space) and less "off-the-shelf" building blocks, the system designer must also be sensitive to the less tangible parameters such as design complexity, testability, and above all cost (for both development and production). Presumably, as more such systems become operational, and all associated technologies become more mature the mathematical relationships of the less tangible parameters will become better understood and less subjective.

Initial system design starts with the desired requirements for the communications link itself: data rate, required bit error rate and range. For the most part, these requirements drive system design in much the same manner as for RF system design. In addition, however, the design of an optical communications system must also take into account the characteristics of the platform on which each terminal is mounted. For the laser system, considerations such as attitude control, and both rigid and nonrigid body on-orbit dynamics will drive the desired transmit beamwidth, and the implementation of link acquisition and pointing control.

2.9.1 Determining The Optimum Transmitter Beamwidth

As discussed, laser communications systems typically operate at or near the diffraction limit of the transmitting telescope. As such, the profile of the control lobe of the beam pattern has a nearly Gaussian dependence with off-axis angle. By taking the derivative of the curve of Figure 2-10 and setting the result to zero, one finds that the optimum transmit beamwidth is given by:

$$\theta_t \simeq 2.838\ \theta_\varepsilon \tag{2-43}$$

where θ_t is the full beam diameter to the e^{-2} points, and θ_ε is the bounded pointing error given by Equation (2-42) (for the pure Gaussian antenna pattern Equation (2-43) becomes an exact expression). Thus, before a transmit beamwidth (hence antenna gain) may be determined some evaluation must be made as to both platform dynamic motion and the design of the pointing control system to correct for that motion. A curve showing the relationship between optimum antenna gain (beamwidth) and system pointing performance is shown in Figure 2-25.

2.9.2 Selecting the Laser

The designer of the spaceborne laser communications systems today has two options for space proven laser sources: the semiconductor laser diode and the solid-state diode-pumped Nd:YAG laser. Systems using the laser diode generally employ direct modulation of the diode drive current and are somewhat limited as to ultimate data rate and range as compared to the diode-pumped Nd:YAG laser-based systems. However, for some applications the advantages in efficiency and weight for laser diodes may outweigh its limitations. A more thorough discussion on the characteristics of laser diodes is contained in Chapter 3.

The diode-pumped Nd:YAG laser, although more complex and heavier than the direct diode source emits a highly collimated beam yielding diffraction-limited performance from the transmitter antenna. As discussed, the Nd:YAG laser may be operated Q-switched, cavity dumped, or mode-locked. The choice is dependent upon data rate and modulation format. In Figure 2-26, the typical operating pulse rates for the various operating modes is shown with their relative peak power advantage. Advantage here is used to mean the ratio of peak output power to average output power. Since average output power is a measure of the overall efficiency for electrical input power, this advantage may be looked upon as a measure of overall system efficiency or relative link margin.

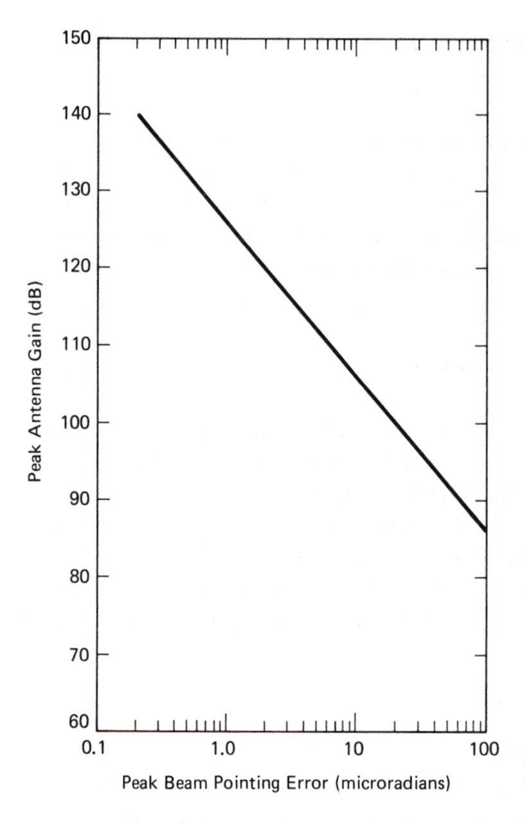

Figure 2-25 Optimum antenna gain versus pointing error.

As seen from Figure 2-26, the output pulse energy is dependent upon pulse repetition rate. To minimize the variation in system operating parameters due to pulse amplitude variations, the modulation ratio (the ratio of the pulse interval modulation to the average pulse interval) is usually limited to 10 percent. Another consideration for system design is the determination of the correct receiver slot width. For Q-switched and cavity dumped lasers typical pulse widths are between 10 ns and 100 ns depending upon specific laser characteristics. The optimum receiver gate width to detect the arriving pulse is on the order of twice the $1/e^2$ pulse width. This provides the minimum sensitivity in the effects of amplitude and timing jitter while highest communications bandwidth. Figure 2-26 can now be used to establish relative transmitter advantage versus data rate for the various laser operating needs.

2.10 AN EXAMPLE OF A LINK DESIGN CALCULATION

After reviewing the various trades for system design, it is considered worth reviewing the entire link allocation process.

As an example, consider a design for a satellite to satellite laser crosslink.

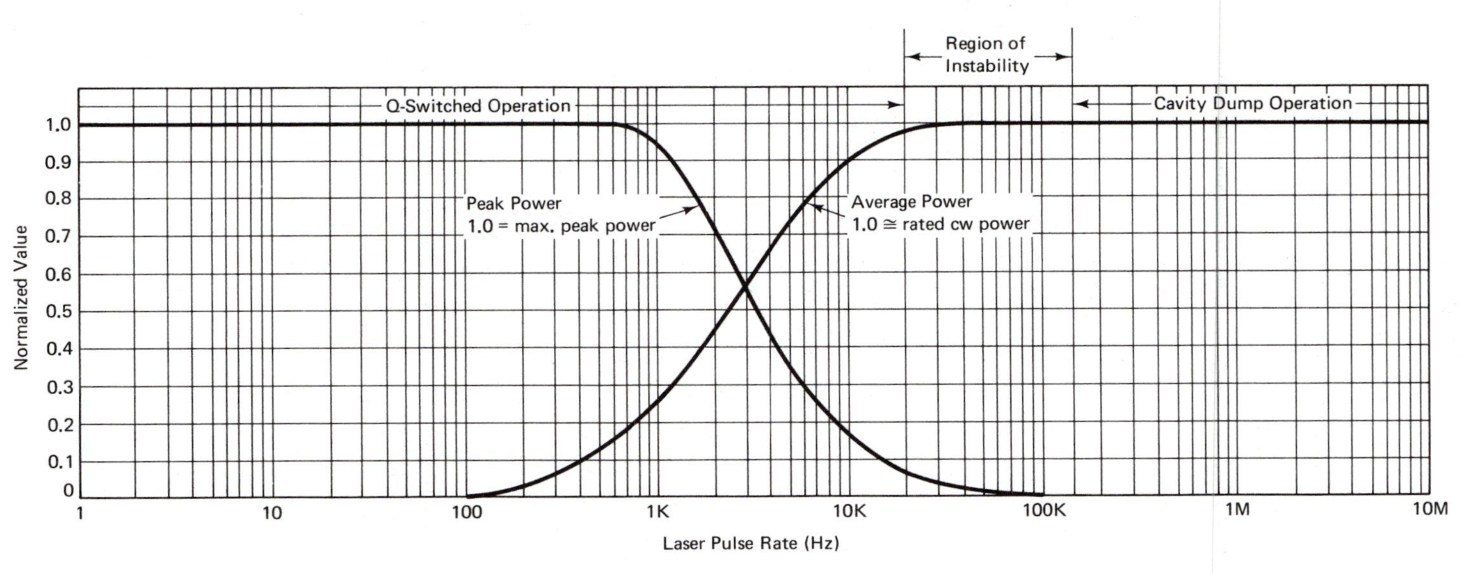

Figure 2-26 Peak and average power versus pulse repetition frequency. Figure shows typical operating rates for Q-switched and cavity pumped Nd:YAG lasers. Typically Q-switched lasers designed to operate below 100 pps utilize both gain switching (flash pumping) as well as Q-switching for improved efficiency. Generally, the bandwidth limit of the intercavity modulator is 10 MHz. Mode-locked operation is used for laser pulse rates between 100 Mpps and 1,000 Mpps.

The user data link is between two geosynchronous satellites stationed in the eastern and western hemispheres, a distance on the order of 84,000 km. The user requirements are to communicate at a bit rate of 1.5 Mbps with a probability of bit error of 1×10^{-6}. The satellite is inertially stabilized but the bounded pointing error from all sources is determined to be on the order of 3.5 μrad. From Figure 2-25, the optimum gain is given. The beamwidth to accommodate the pointing jitter is 10 μrad (for a diffraction-limited Cassegrainean aperture of 190 mm). For optimized Cassegrainean geometry, the theoretical limit for on-axis far field gain is 112 dB.

From Figure 2-26, the laser operating mode with natural repetition rates near the data rate is a Nd:YAG laser operating in the cavity dumped mode. To achieve the required data rate an *m*-ary pulse interval modulation format should be used. For this design example it can be assumed that a typical cavity dumped laser will be used with pulsewidths of 10 ns (FW/e^2). To evaluate which order modulation is optimum, a table of modulation parameters can be constructed (see Table 2-1).

Operating cavity-dumped lasers beyond 500 kpps repetition frequency creates design difficulties for cavity-dumped laser modes in addition to the overall inefficiency of the laser. The high pulse-to-pulse amplitude variations resulting from modulation depths >50 percent adversely drive the complexity of the communications receiver. Thus, the options of 4, 5, or 6 bits per pulse appear to be the most viable formats for this link example. The system designer must choose between the design impact of accommodating the pulse-to-pulse amplitude variations for the high percent modulation or make up the loss in efficiency in the system. For the purposes of this example, 6 bits per pulse (64 slots) are selected with a laser repetition rate of 250 kpps. Typical laser operating parameters at this repetition rate are 60 Watts electrical pump power, 150 mW average output, 60 W peak, 10-ns pulsewidth.

Having established the modulation format and laser-operating parameters, there remains only to determine the required signal at the detector for desired bit error rate and the receive antenna gain. Almost all laser communications use avalanche photodiodes. Today's state-of-the-art APDs tout >35 percent quantum efficiency at 1.06 μm and <0.01 ionization coefficient.

TABLE 2-1 MODULATION FORMAT PARAMETERS

Number of Bits/Pulse	*Laser Repetition Frequency*	*% Modulation*	*Link Efficiency*
2	750 kpps	6	−6.0 dB
3	500 kpps	8	−4.2 dB
4	375 kpps	12	−3.0 dB
5	300 kpps	20	−2.0 dB
6	250 kpps	32	−1.2 dB
7	214 kpps	55	−0.6 dB
8	188 kpps	86	0 dB

TABLE 2-2

Laser peak power	16.0 dB
Transmit optics efficiency	−3.0 dB
Transmit antenna gain	112.3 dB
Wavefront errors	−1.0 dB (λ/12 error)
Pointing loss	−4.4 dB
Propagation loss	−300 dB
Receive antenna gain	115 dB
Receive optics efficiency	−3.0 dB
Required signal	−74.6 dB
Margin	6.5 dB

Applying the equations for APDs and bit error rate for PIM developed earlier, one finds that the typical required signal at the receiver is 60 nW peak without any significant optical background sources.

Having established minimum beamwidth (transmit antenna gain), laser output power and required signal, the last step is to determine the required receive antenna gain. The laser crosslink systems usually operate as transceiver: simultaneously transmitting and receiving through the same telescope antenna. Thus, the receiver gain geometry is the same as the transmitter geometry and gain is computed using Equation (2-9).

A first cut link table can be constructed based on these initial allocations. Table 2-2 shows the link tabulation.

The system designer is starting with a comfortable link margin some of which can be relaxed to alleviate some system constraints. Most satellite communications links operate with as low as 3.0-dB margin to preserve performance.

Close examination of the link table reveals what options at this point are available to the system designer for performance allocations. The propagation loss is a fixed physical parameter and is traded only against desired user range. Laser peak power is strongly a function of available prime power from the platform. Technical enhancements of a specific laser type usually cannot be expected to yield significant improvements in link performance. Most optical communications subsystems require complex optics trains involving as many as 50 to 100 surfaces in an optical path. Even to achieve the total transmission of 50 percent shown in the link table requires high-performance optical coatings. Again, further technical improvements will not yield significantly improved link performance (unless the total number of surfaces can be reduced). A strong driver on link performance, however, is transmitter pointing. The assumption of 3.5 μrad of point error is based on a budget driven primarily by static coalignment biases (and only to a small part by angle processor random noise). Improving alignment techniques, structural stability, and thermal stability could significantly reduce pointing errors. This has the double effect of permitting both increased transmit antenna gain and receive antenna gain. The cost for this improvement, however, is some increase in package weight.

2.11 AN EXPERIMENTAL AIRCRAFT TO GROUND SYSTEM

The Laser Communications Airborne Flight Test System (AFTS) program was directed toward the development and evaluation of critical components and design concepts applicable to a high data-rate spaceborne laser communications systems using the Nd:YAG laser. These concepts were demonstrated over an experimental link between an aircraft and ground station receiver (Figure 2-27). The 1977–1980 effort represented an intermediate milestone for future space applications. The technology developed served as the foundation for current operational systems.

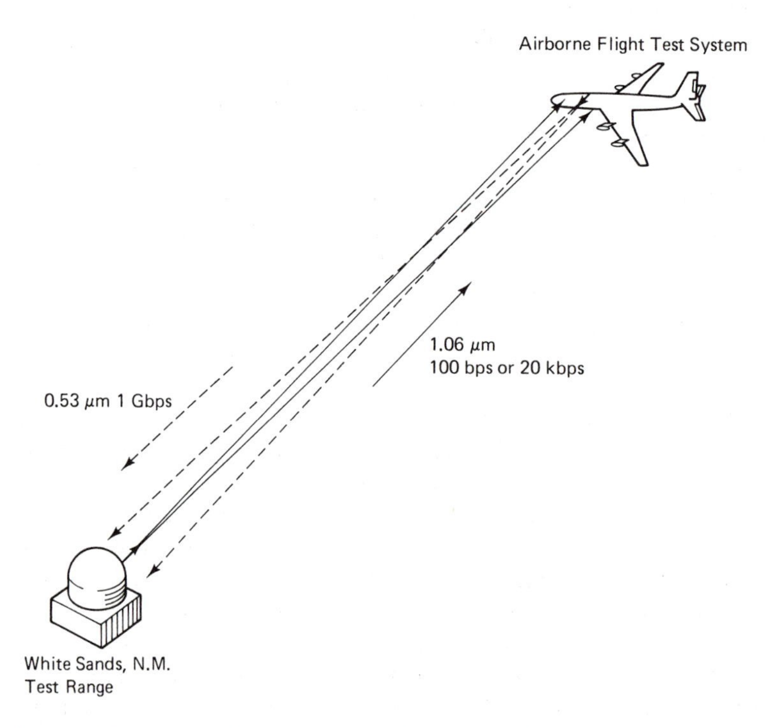

Figure 2-27 Airborne Flight Test System Experiment: Experiment conducted in 1980 was successful in demonstrating the feasibility of communicating with the Nd:YAG laser. Several different modulation formats were used to transmit data between aircraft and ground. During communications the aircraft terminal transmitted a 100-μrad divergence beam, and the ground station transmitted a 1 mrad uplink beam. Downlink data included several color TV signals and audio. Audio and teletype data was transmitted over the optical uplink.

The primary objectives of the program were to:

1. Demonstrate the transmission of data via an open laser beam from one location to another
2. Demonstrate that acquisition and tracking can be accomplished with narrow laser beams
3. Close an optical communications link with hardware performance representative of a spaceborne system

The communication, acquisition, and tracking functions were accomplished by placing a high data-rate laser transmitter package in a USAF KC-135 (Figure 2-28) and acquiring, tracking, and communicating with a receiver package located in an optical ground station in White Sands Missile Range in New Mexico (Figure 2-29).

The hardware performance allocations were based on what would be required of a spaceborne laser communications terminal similar to those just described. In fact, in certain respects, operation of the link over the 100-km

Figure 2-28 Aircraft test bed carried the high data-rate transmitter package; the electro-optic components were built to spaceborne prototype designs and performance requirements. The aircraft was modified to add a high-quality optical window through which the aircraft terminal could transmit and receive the optical information. The white truck contains optical equipment used to check-out the aircraft terminal prior to each flight. (Courtesy McDonnell-Douglas)

Figure 2-29 AFTS ground station: The dome building contained the special laboratory hardware necessary for transmitting to and receiving optical data from the aircraft. A two-axis gimballed mirror contained in the dome was capable of hemispherical coverage to accommodate aircraft flight profiles. (Courtesy McDonnell-Douglas)

atmospheric slant path placed more severe constraints on the system performance than would be required for a typical space link of 80,000 km. The design of the electro-optic components (laser, optics, telescope, modulator, detectors) was intended to be as near to space prototype as practical, while the associated electronics and ground station were built using conventional components.

During the experiment, the airplane flew a circular trajectory around the ground station at slant ranges from 10 km to 100 km. The high data-rate transmitter acquired and tracked a 1.06-μm beacon to obtain line of sight pointing information for the high data-rate transmitter. The airborne high data-rate transmitter transmitted a narrow beam (<100 μrad) providing a data-rate of 1,000 Mbps. Instrumentation was provided on the aircraft and at the ground station to measure acquisition and tracking times and accuracies, communications bit-error rates, determine slant path atmospheric effects and make direct measurement of far field patterns via scanning of the transmit beam. The 1.06-μm beacon provided 100 bps and 20 kbps optical uplinks using pulse-interval modulation.

2.11.1 Flight Hardware

The flight hardware was divided into an electro-optic assembly (shown in Figure 2-30) and an electronic assembly (Figure 2-31). The system block diagram in Figure 2-32 shows the key elements of this system. The Electro-

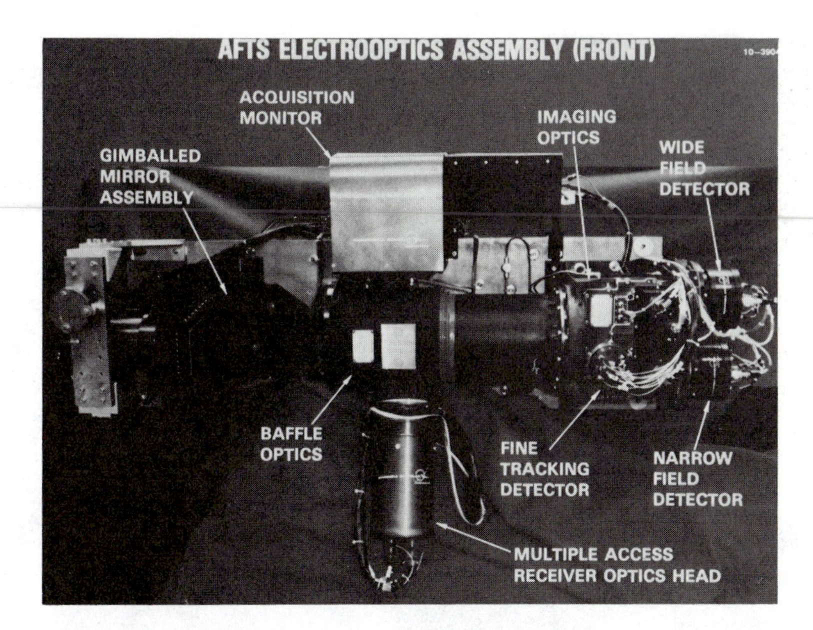

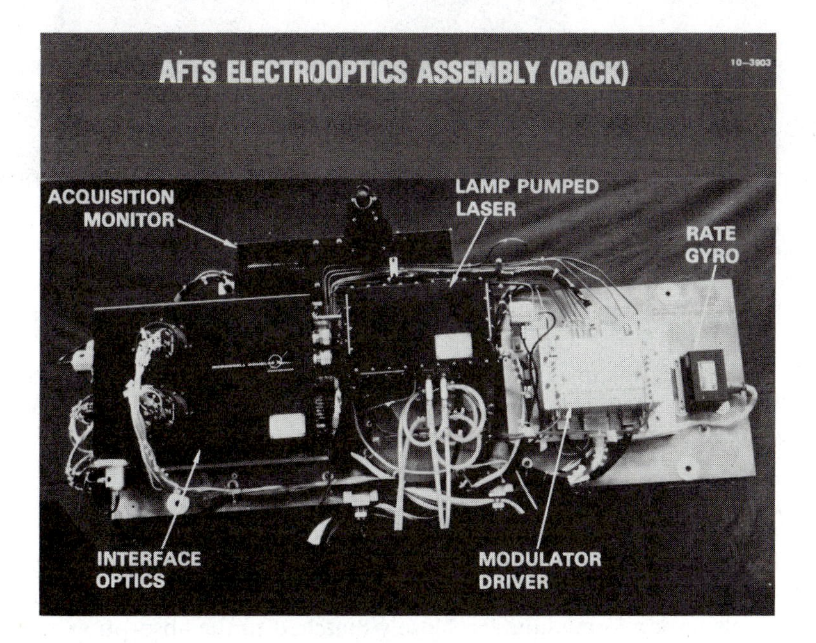

Figure 2-30 Aircraft terminal electro-optics package: Supported on a balanced single-point mount for isolation from aircraft motion, all electro-optic components were attached to a rigid baseplate. The gimballed mirror provided coarse pointing of the 5 μrad, 7-1/2-in. diameter beam. The mode-locked and frequency doubled Nd:YAG laser produced 180 mW of average power that provided ample power at the ground terminal for maintaining communications even in the presence of atmospheric turbulence. (Courtesy McDonnell-Douglas)

Figure 2-31 To maintain flexibility and support diagnostic activities, all the flight terminal electronics included operator consoles and instrumentation. Once equipment was configured for a particular experiment, operation of the airborne terminal was autonomous.

Optic Assembly was installed in an environmental enclosure, which was located behind a thirty-inch optical window. Electronic subassemblies and instrumentation were located in racks with operator consoles.

High data-rate laser

The AFTS lamp-pumped laser was a mode-locked, frequency doubled Nd:YAG laser. This laser was efficiently pumped by a rare earth metal arc lamp whose spectral emission was closely matched to the absorption spectra of the neodymium-doped rod.

The laser utilized a folded cavity design for compactness and improved doubling efficiency with a single intracavity device used for acousto-optic mode locking and frequency doubling. This device was a $BaNaNb_3O_{15}$ crystal driven with an acoustic transducer at 250 MHz and thermally stabilized at

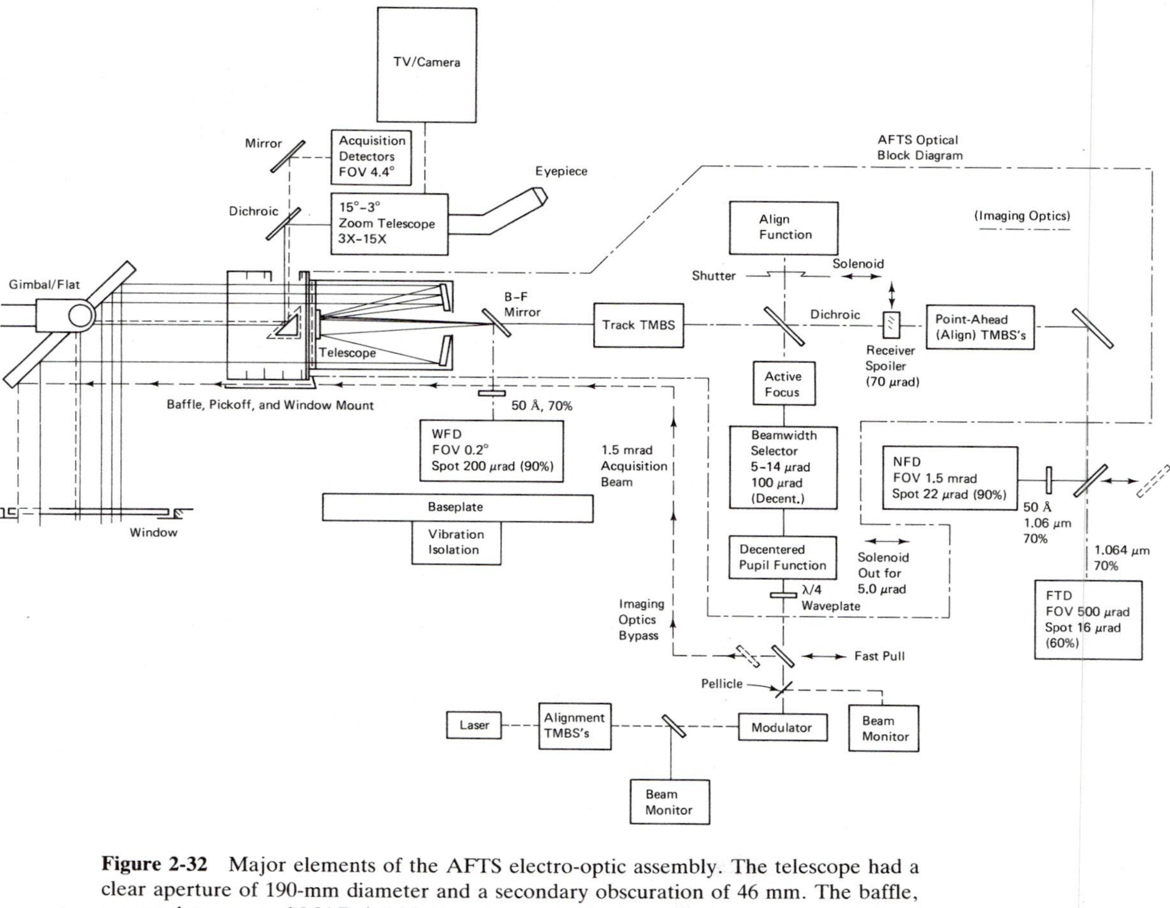

Figure 2-32 Major elements of the AFTS electro-optic assembly. The telescope had a clear aperture of 190-mm diameter and a secondary obscuration of 46 mm. The baffle, zoom telescope, and MAR (multiple access receiver) were added to the assembly for the airborne experiment. The narrow field detector and fine tracking detector both performed the same function. Both were included to perform detector technology evaluations.

the phase-matching temperature. Typical operating performance of the laser during the field operations was 180 mW average output power for 250 W of prime pump power and with a 350-ps pulse width. These output parameters are the same as or better than would be required for a space application, although laser diode pumping has significantly reduced the prime pump power to less than 60 W for the same output parameters.

High data-rate modulator

The AFTS High Data-Rate Modulator used a combination of polarization binary modulation and time-delay binary modulation collectively referred to as *pulse quaternary modulation* (PQM) to achieve a 1-Gbps channel bandwidth from a 500-Mpps, fixed-repetition-rate laser source. The modulator consisted of two separate polarization modulators. Time-delay modulation was accomplished by selectively passing the optical pulse through a 1-ns passive time-delay unit. The output then passed through a second polarization modulator. Each modulator section used lithium tantalate crystals driven with a 22-$V_{p\text{-}p}$ waveform. In addition to the full 1-Gbps channel capacity each modulator could be driven separately for 500 Mbps PDBM or PPBM modulation formats.

The measured transmission for the entire modulator assembly on AFTS was 70 percent.

Imaging optics assembly

The AFTS Imaging Optics Assembly (IOA) performed the functions of receiving the 1.06 μm beacon illumination and relaying it through various optical paths to one of the acquisition or tracking detectors. It also shaped and directed the 0.532-μm transmit beam down a coincident optical path. Depending on the mode selected, the IOA could transmit either a diffraction-limited 5-μrad beam full aperture, or a decentered, 100-μrad beam. An automatic static alignment path was included to co-align the noncommon portions of the transmit and receive optics. Although not required on the airborne experiment, a "point-ahead" function was also included to introduce a carefully calibrated pointing bias between the transmit and receive functions to correct for on orbit velocity of light aberration.

Several distinct receive paths were incorporated into the IOA for acquisition and tracking of the 1.06-μm beacon. The wide field acquisition detector (WFD) field of view is $\pm 0.15°$. This is the field required to cover the initial pointing uncertainties of a typical geosynchronous satellite pointing to another geosynchronous satellite or a ground terminal. The narrow field acquisition detector (NFD) was the second detector used in the acquisition and tracking to further reduce optical pointing errors and maintain fine tracking. Separation between the WFD and NFD viewfields is accomplished by means of a bifurcating mirror at the telescope image plane. Both receive paths relayed the incoming beacon signal to quadrant arrays of avalanche photodiodes for processing of angle information. Due to the sensitivity of the APD to optical background illumination a 50-Å bandwidth interference filter was included to reject a significant amount of optical background.

Precision 300-Hz tracking loops were required to support the pointing of the 5-μrad beam in the presence of typical spacecraft dynamics. Special beam steering devices were developed. These were small, light-weight torque motor devices which drove low inertia fold mirrors. These torque motor-driven beam steerers (TMBS) were equipped with precision position feedback for tracking loop compensation. In laboratory testing, random angle processor noise of less than 0.6 μrad, 3σ was obtained with the system.

The Cassegrain telescope clear aperture was 0.191 m and the central obscuration 0.046 m. To support the required antenna gain, an extremely precise wavefront quality was required from the telescope. The measured value for this telescope was $\lambda/50$ at 0.532 μm, making it one of the highest quality small aperture Cassegrain telescopes ever built. The construction was of Invar spider and barrel and special low coefficient of expansion glass optics. Internal knife-edge baffling with special absorbing anodizing over the complete interior surface was added to minimize stray light scatter. Consistent with spaceborne system requirements, the measured telescope wavefront quality contributed less than 0.07 dB of on-axis loss in the diffraction-limited, 5-μrad transmit beam. The entire imaging optics assembly including the telescope had at beginning of life better than 70 percent transmission on the 5-μrad transmit path, and better than 68 percent transmission on the receive path.

Acquisition and tracking detectors

To perform the functions of acquisition and tracking with the 1.064-μm ground station beacon, the AFTS used a converging acquisition and tracking handover sequence that employed the wide field detector (WFD) and narrow field detector (NFD). Using a technique analogous to the monopulse processor in conventional RF radar systems, each of these assemblies used a quadrant array of detectors ratiometrically processed to derive very precise angle information. To pass the narrow (100 ns) pulses from the Q-switched beacon laser, the detectors had an electrical bandwidth of 3.5 MHz. The performance of the assemblies was driven by maximum sensitivity for the acquisition detectors, and by accuracy and stability for the tracking detector.

APDs were developed with very low ionization coefficients. The devices were made thick to maximize the quantum efficiency in the transparent silicon. To compensate for the thickness, the active area was reduced to reduce bulk noise and device capacitance. The result was a low noise device of 35 percent quantum efficiency at 1.064 μm and usable at high avalanche gains. These devices demonstrated the required detection performance (90 percent detection probability and unity false alarm rate) with less than 2 nW of peak pulse power. Results from the AFTS tests showed that the quadrant array of APDs not only had the sensitivity required for acquisition but also had the accuracy and stability required for fine tracking.

2.11.2 Tracking Ground Station

The tracking ground station (block diagram shown in Figure 2-33) contained the hardware for optically acquiring and tracking the airborne transmitter, transmitting the 1.064-μm beacon signal for the transmitter terminal acquisition and tracking functions, and the special, high-speed detectors for receiving, detecting, and processing the high data-rate downlink information. In addition, special processing electronics were added to evaluate the quality of the high data-rate information and automatically acquire both data quality and received signal strength data for performance analysis. The facility was equipped with a steerable dome and two axes gimballed mirror and special "clean" laboratory space for the receiver terminal itself.

Gimballed mirror

The gimbal was constructed and installed in the steerable dome of the ground station. The two-axis gimballed mirrors could steer a beam from some point on a hemisphere down the concrete pedestal to a stationary fold mirror and into the receiver terminal. The baseplate turned about a vertical axis along the centerline of the pedestal. The azimuth mirror folded the beam along that axis horizontal into the elevation mirror along its axis of rotation. The elevation mirror folded the beam 90 to a line from the horizon to zenith depending upon the angular position. Gimbal control was achieved using both rate and position feedback from each gimbal axis.

The gimbal initial pointing commands were computed from radar data providing the aircraft location from the precision tracking radar.

Receiver terminal

The receiver terminal performed all of the functions that would be required of an operational receiver. The block diagram in Figure 2-33 shows the functional arrangement of the various elements of the terminal.

Optical acquisition and precision tracking of the high data-rate transmitter was achieved using a special scanning photomultiplier tube to provide angle of arrival data to the fine tracking torque motor-driven beam steerers (TMBSs). The detector was scaled to cover a 0.25° field of view for acquisition and 20 μrad viewfield in the fine tracking mode. Using this approach, an optical tracking control bandwidth of better than 150 Hz and tracking accuracies of better than 4 μrad, rms was achieved during flight tests. To accommodate the atmospherically induced scintillations a dual mode automatic gain

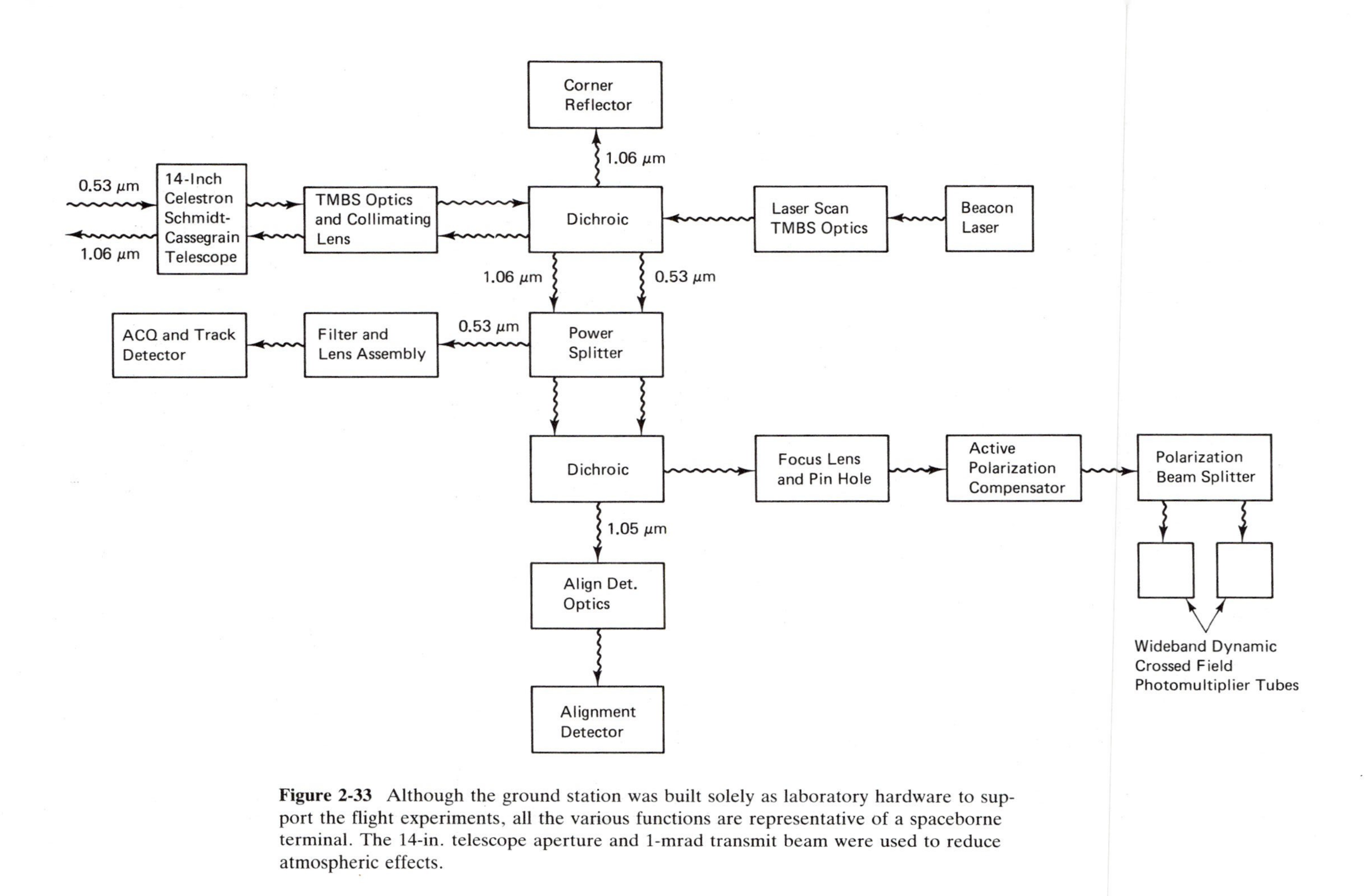

Figure 2-33 Although the ground station was built solely as laboratory hardware to support the flight experiments, all the various functions are representative of a spaceborne terminal. The 14-in. telescope aperture and 1-mrad transmit beam were used to reduce atmospheric effects.

control was incorporated providing 20 dB of control in a 1-KHz bandwidth and 60 dB of control in a 1-Hz bandwidth.

Beacon illumination of the aircraft was obtained with a Q-switched 1.064-μm Nd:YAG laser. To cover the initial acquisition uncertainty area, the 3,000 pps laser was scanned over the 0.25° telescope viewfield at a 10 frames/s rate. To provide adequate overlap, the scan optics were adjusted to achieve a full width divergence of 1 mrad. As the acquisition process progressed, and illumination was received from the aircraft, the scan coverage was decreased and frame rate increased to provide increased illumination to the aircraft. With transition to fine tracking, the scanning was stopped, and the beam narrowed to a 100-μrad divergence, to provide sufficient illumination to the aircraft for transition to fine tracking. Communications electronics was included in the Q-switch drive electronics to transmit up to 20 kbps of data over the beacon uplink. This was achieved by pulse interval modulation (PIM) of the beacon laser pulse stream.

The high data-rate receiver consisted of a pair of high-speed gated photomultiplier tubes known as dynamic crossed field photomultipliers (DCFPs), which provided high gain and bandwidth with extremely low excess noise factors. These detectors were located behind a polarization beam splitter to separate the polarization encoded data. The field of view of the detectors was limited to <100 μrad by a pinhole in the optical path. Background noise was further limited by a 25 Å filter in front of each detector. With the limited field of view, narrow passband, and receiver gating, the detectors were capable of the same quantum limited performance independent of day or night illumination. To compensate for ellipticity in the circularly polarized aircraft transmit beam, a variable polarization compensator was included in the optical path to adjust the relative strength of the two polarizations.

Alignment of the 1.06-μm transmit path and of the 0.532-μm receive paths was achieved by retroreflecting a portion of the parasitic leakage in the diplexer/dichroic to a precisely aligned alignment detector. Error signals developed this way were fed to the scan TMBS to correct boresight misalignments.

AFTS flight tests

During three months of flight testing over 35 flights were flown totalling better than 200 hours of experiments. Specific tests were performed on the system to evaluate the performance of the initial acquisition sequence, the acquisition handover process, and fine tracking performance. Those tests included evaluation of the system as designed and parametric evaluation of system design by varying parameters such as control bandwidth, and signal level. Performance data was also taken on both the high data rate down link and the 20 kbps uplink through evaluation of bit data rate on PN data and by a subjective evaluation of voice and video data being transmitted over the link. Amplitude scintillation and angle of arrival data was taken at both terminals for characterization of the atmospheric channel.

The results of the flight testing confirmed the feasibility of high data-rate communications with the Nd:YAG laser. The ability of the space design equipment to communicate within expected bounds through the atmosphere demonstrated the capability of the hardware design and has lead the way to operational spaceborne laser communications systems.

REFERENCES

1. M. Ross, et al., "Space Optical Communications With The Nd:Yag Laser." *Proc. IEEE*, Vol. 66, No. 3, March 1978, pp. 319–45.
2. There are a number of texts that describe laser principles and general operating characteristics. A good layman's introduction to lasers and laser terminology can be found in D. C. O'Shea, et al., *Introduction to Lasers and Their Applications*, Edison Wesley, Boston: MA, 1977.
3. J. H. McElroy et al., "CO_2 Laser Communications Systems for Near-Earth Space Applications," *Proc. IEEE*, Vol. 65, No. 2, Feb. 1977, pp. 221–51.
4. M. Ross et al., *ibid.*
5. A. Yariv, *Introduction to Optical Electronics* (New York: Holt, Rinehart, and Winston, Inc., 1971), pp. 177–221.
6. A. Yariv, *Ibid.*, pp. 222–46.
7. D. Maydan and R. B. Chester, "Q-Switching and Cavity Dumping of Nd: YAG Lasers," *J. Applied Physics*, Vol. 42, No. 3, March 1971, pp. 1031–34.
8. R. B. Chesler and D. Maydan, "Calculation of Nd: YAG Cavity Dumping," *J. Applied Physics*, Vol. 42, No. 3, March 1971, pp 1028–30.
9. D.T. Paris and F. K. Hurd, *Basic Electromagnetic Theory* (New York: McGraw-Hill Book Co., 1969), pp. 458–510.
10. D.T. Paris, *Ibid.*
11. J. W. Goodman, *Introduction to Fourier Optics*, (New York: McGraw-Hill Book Co., 1968), pp. 57–74.
12. J. H. Oberhauser, "Optical System Approach to Cassegrain Telescope Selection, Design, and Tolerancing," *SPIE*, Vol. 193, *Optical Systems Engineering*, pp. 27–33.
13. P. P. Webb, R. J. McIntyre and J. Conradi, "Properties of Avalanche Photodiodes," *RCA Review*, Vol. 35, June 1924, pp. 234–277.
14. M. Ross, et al., "Space Optical Communications with Nd:YAG Laser." *Proc. IEEE*, Vol. 66, No. 3, March 1978, pp. 342–43. ©IEEE.

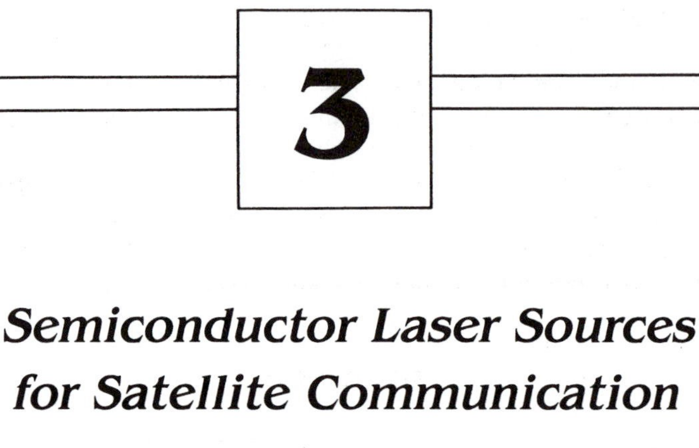

3 Semiconductor Laser Sources for Satellite Communication

Dr. Gary Evans and Dr. Michael Ettenberg
RCA Laboratories
Princeton, NJ 08540

3.1 INTRODUCTION

Semiconductor lasers are the ideal light source for optical transmitters in space communications due to their small size and weight, high efficiency and reliability. In addition, semiconductor lasers are easily modulated by direct current injection.

The laser chip size is approximately 100 × 200 × 200 μm with most of the weight and size being taken up by the Cu heat sink as illustrated in Figure 3-1. Selected "champion" devices have produced about 200 mW cw[1,2] in a single device and about 2.5 W[3] in a multimode array. The light output to electrical input conversion efficiency is as high as 35 percent[4,5] making semiconductor lasers the highest efficiency laser source. The median operating lifetime of semiconductor lasers has been measured to be on the order of 10^5 to 10^6 h[6,7,8] and diode lasers have been directly modulated at room temperature at rates of up to 12 GHz.[9,10] The problem for space communications applications is that each of the operating limits discussed above have been achieved with specially designed devices that are lacking in other important characteristics, or devices operated beyond their reliable operating limits. The present limits described above set the targets for laser design and provide the limits for trade-offs in communication system use. It is probable that with the present pace of semiconductor laser technology development, reliable and useful devices with all the forementioned attributes available simultaneously will be developed in the next three to five years. In this chapter, we will describe in detail some of the important structural and operating characteristics of semiconductor lasers and their impact on space communications systems.

The present state of technology requires the combined output of several laser diodes to obtain the necessary average power for high data-rate (several hundred Mbits/s or more) space communications. The outputs can be combined in three basic ways: (1) incoherently via conventional optics; (2) coherently using integrated phase-locked arrays or master-slave coherent locking concepts; or (3) as an optical pump replacing lamps for a solid state laser. In all schemes using semiconductor lasers for space communications, changes with aging in such operating characteristics as lasing spectra, far-field radiation

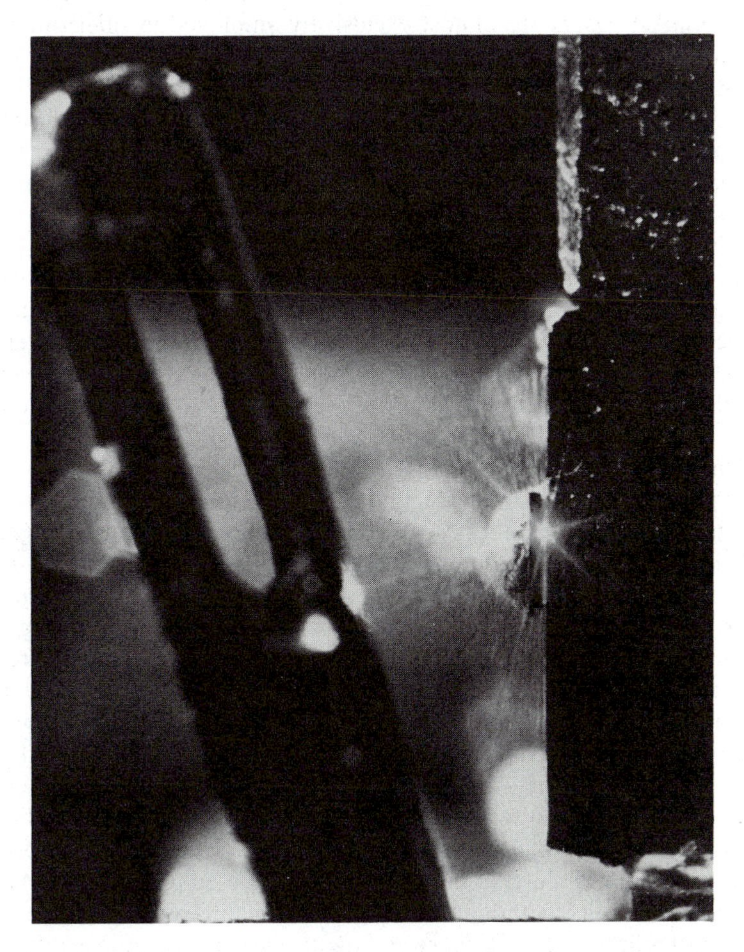

Figure 3-1 Visible cw room temperature laser diode and an eye of a small needle.

patterns and threshold current can limit the laser communication system to a lifetime much less than the 10^5 hours or more reported lifetime of individual state-of-the-art semiconductor lasers.

Today's semiconductor lasers are based on two semiconductor materials systems, AlGaAs and InGaAsP. These III-V compounds can be lattice-parameter matched to GaAs and InP substrates respectively. In this way, the necessary multi-alloy-layer structures can be grown without the introduction of defects. The AlGaAs lasers can be made to emit reliably between 0.78 and 0.86 μm and the InGaAsP lasers between 1.2 and 1.65 μm. The AlGaAs lasers, with their shorter emission wavelength, are the desirable source for optical systems requiring a small far-field spot size, small beam divergence, and high photon energy. These lasers are presently employed commercially for optical disc recorders, disc players, and laser printing. The longer wavelength InGaAsP lasers are almost exclusively employed in fiber optic communications due to the lower loss and lower chromatic dispersion in these fibers at the longer wavelengths. In addition to size, weight, reliability and efficiency, the following characteristics make AlGaAs diodes desirable for space applications: (1) a smaller beam divergence for fixed optics (or the use of smaller optics for fixed-beam divergence); (2) the emission wavelength of AlGaAs lasers (ranging from 0.8 to 0.9) is an excellent match to silicon avalanche photodetectors (APD); and (3) the demonstration of output powers in excess of 100 mW. Silicon avalanche detectors are uniquely sensitive due to a very high electron to hole ionization ratio not found in Ge or other III-V semiconductor compounds that are sensitive at wavelengths longer than 0.9 μm. In addition, longer wavelength detection requires smaller bandgap materials, which in turn mean higher leakage current and resulting increased noise.

However, in the future, the use of quaternary GaInAsP lasers cannot be ruled out. Although they are undesirable because they have a longer emission wavelength (1.3 to 1.5 μm) resulting in a weight and cost penalty associated with the larger emitter and receiver optics, three important reasons make them strong contenders for space communications: (1) They provide the best match for low loss and dispersion in optical fibers, causing a tremendous amount of development work in quaternary lasers—therefore, they will be the first semiconductor lasers to be commercially available with dynamic single mode stabilization; (2) unlike ternary semiconductor lasers, the quaternary lasers are not limited in power output by the peak optical power density at the output mirror facet—ultimately, discrete quaternary lasers may be able to deliver two to ten times the power of ternary devices; and (3) the technology for detectors in the 1.3 to 1.5 μm range is continually improving.

The data and discussion in this chapter are based primarily on AlGaAs semiconductor lasers, although when there is a difference in some of the characteristics between the ternary and quaternary devices, those differences are discussed.

3.2 BACKGROUND

In its simplest embodiment we can consider a laser diode to be a three-dimensional box in which electrons and holes are injected to invert the population. As illustrated in Figure 3-2, the walls of the box are the facet mirrors (longitudinal direction), heterojunction layers (transverse direction) and stripe contact (lateral direction). When a voltage is applied to the diode in the forward direction (+ to the *p*-side), little current is drawn until the energy bands are flattened, as illustrated in Figure 3-3, which is an energy band diagram of an (AlGa)As double-heterojunction laser. The voltage (V) needed to flatten the energy bands corresponds approximately to the bandgap energy eV of the middle layer or active region. As the voltage is increased beyond the "knee" or energy band flattening voltage the current flow is limited only

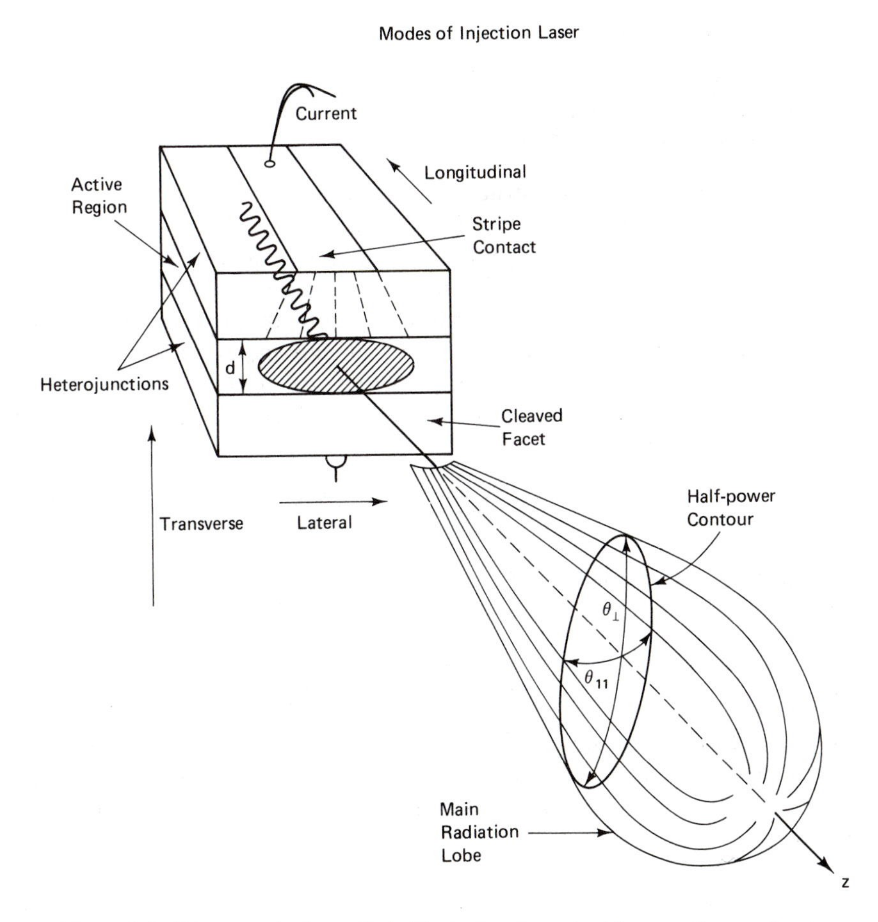

Figure 3-2 Diagram of a laser diode illustrating the output beam pattern and the modes of a laser diode.

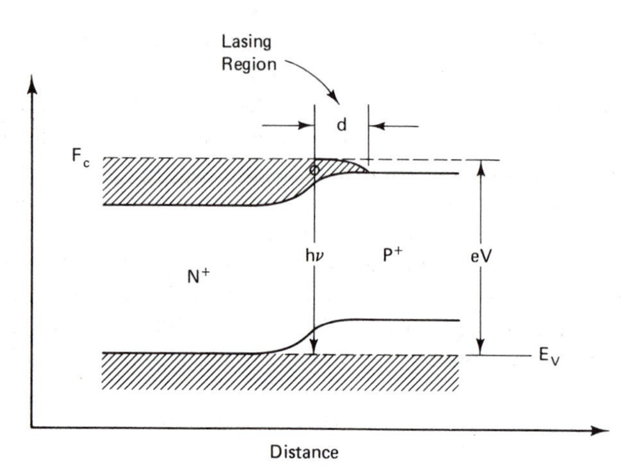

Figure 3-3 Band diagram of a forward-biased laser diode with applied voltage V and bandgap energy of approximately eV.

by the resistance of the laser structure (typically 1 to 5 Ω). In this regime, as illustrated in Figure 3-4, which shows the light output versus current input (*L-I*) and current input versus applied voltage (*I-V*) characteristics of a typical laser diode, the light output efficiency is low until threshold for lasing is achieved (enough carriers are introduced to provide population inversion) and then the light output efficiency is dramatically increased.

The external differential quantum efficiency (η_q), is the incremental number of photons emitted from the laser divided by the incremental number of injected electrons:

$$\eta_q = \frac{\Delta \text{ photons emitted}}{\Delta \text{ electrons injected}} \tag{3-1}$$

The external differential quantum efficiency is related to the slope of the power-current curve by:

$$\eta_q = \frac{\Delta \text{ photons/s}}{\Delta \text{ electrons/s}} = \frac{\Delta P_o}{\Delta I \, h\nu} \tag{3-2}$$

where h is Planck's constant, ν is the photon frequency and ΔP_o is the incremental increase in output power corresponding to an incremental increase in input current ΔI. For the lasing wavelength in angstroms, Equation (3-2) becomes

$$\eta_q = \frac{\Delta P_o}{\Delta I} \frac{\lambda}{12{,}400} \tag{3-3}$$

Many system engineers commonly refer to the incremental slope of the power-current curve in mW/mA as the "slope efficiency" of the device. From Equation (3-2) the "slope efficiency" and the external quantum efficiency are related by the factor $h\nu$, the energy per photon.

Common practice is to refer to the "slope efficiency" and external quantum efficiency per facet. By increasing the reflectivity of the rear facet, R_r, and lowering the reflectivity of the front facet, R_f, with dielectric coatings, the front facet efficiency can be increased by about a factor of two. Changes in facet reflectivity can also effect the threshold current as shown by the relation[11,12]

$$J_{th} = J_t + \frac{1}{\Gamma A}\alpha + \frac{\ln(1/R_f R_r)}{2L} \tag{3-4}$$

where J_t is the current density for zero gain, A is a constant and of the order of 0.035 $cm^{-1}/\text{Å}\ cm^{-2}\ \mu m^{-1}$, L is the length of the laser, and Γ, known as the confinement factor, is the ratio of the mode energy in the active layer to the total mode energy. The quantum efficiency below the lasing threshold is

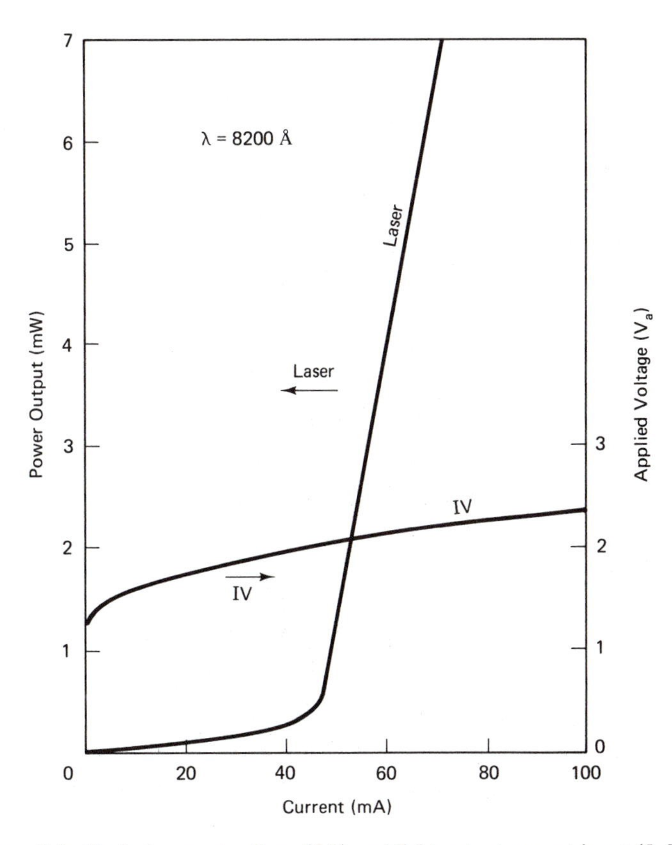

Figure 3-4 Typical current-voltage (I-V) and light output-current input (L-I) characteristics of a laser diode.

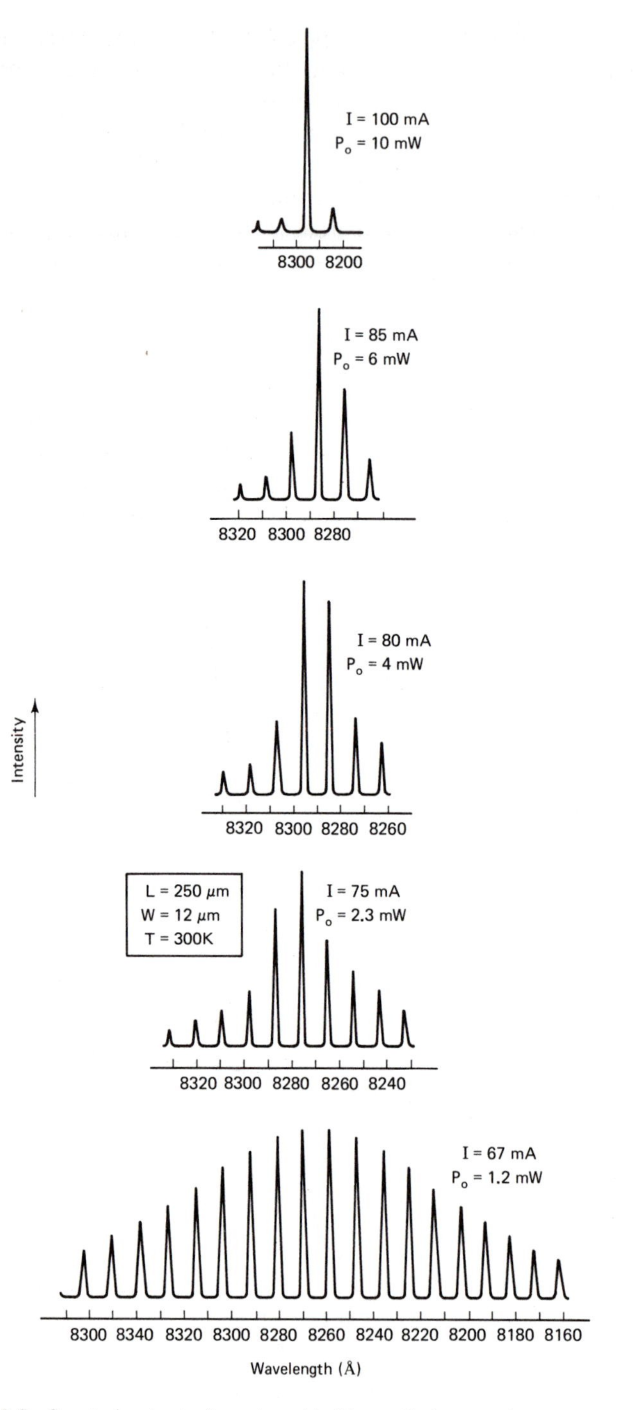

Figure 3-5 Spectral output of a gain-guided laser diode at various power output levels, illustrating discrete spectral lines.

usually less than 1 percent because the light is emitted incoherently in all directions and most of it is absorbed internally. Additionally, since the recombination lifetime is a few nanoseconds, nonradiative processes can also compete for the injected electrons or holes. Above the lasing threshold, the light is almost all directed at the mirrors, the stimulated lifetime is very short (picoseconds) and the external differential quantum efficiency usually is between 20 to 70 percent. J_{th} for modern high-power lasers is between 500 and 2,500 A/cm^2.

The total power efficiency (η_T) is defined as power out of the diode divided by the power input

$$\eta_T \equiv \frac{\text{mW (output)}}{\text{mA (V) (input)}} \tag{3-5}$$

where V is the voltage applied to the diode ($V \sim E_g + IR$). The power efficiency is a strong function of both the diode resistance (R) and the drive current (I). In the reliable operating range of good high-power lasers, the power efficiency is 10 to 20 percent. For systems applications, the important parameter is the power efficiency out of a single facet.

The light output characteristics are shaped by the semiconductor box. Since we want low threshold current, the volume to be inverted is designed to be quite small and the resulting radiation pattern can diverge as much as 50° in both the lateral and transverse directions due to diffraction. The shape of the beam output is fixed by the standing electromagnetic light waves set up by the exact shape and spacing of the transverse and lateral waveguide walls. The spectral output of the laser is fixed roughly by the bandgap of the active region, which determines the approximate wavelength. In addition, the Fabry-Perot modes (the standing wave pattern set up between the laser mirrors) cause the laser to emit in discrete spectral lines (Figure 3-5).

3.3 PERFORMANCE AND GEOMETRIES

3.3.1 Gain-guided Lasers

Until about 1980, most commercially available semiconductor lasers were gain-guided lasers of either the oxide-defined stripe-geometry double-heterostructure (DH) construction shown in Figure 3-6a, or the proton bombarded stripe-geometry DH construction shown in Figure 3-6b. Both of these planar stripe geometries depend on the width of the stripe to provide lateral control of the lasing regions. If the stripe is too wide, multiple filaments within the stripe lase independently. A single lateral (parallel to the junction) mode (roughly Gaussian-shaped) usually results for very narrow stripes (2 to 5 μm). A two-lobed lateral far-field may develop for larger stripes. The oxide-defined and proton-bombarded stripe geometry semiconductor lasers depend on the

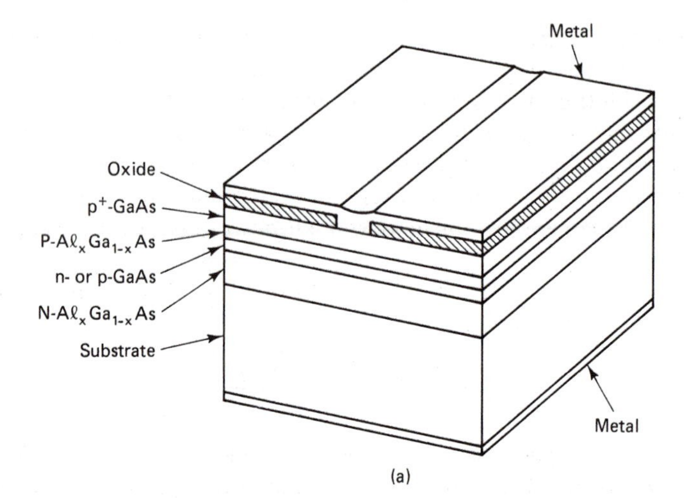

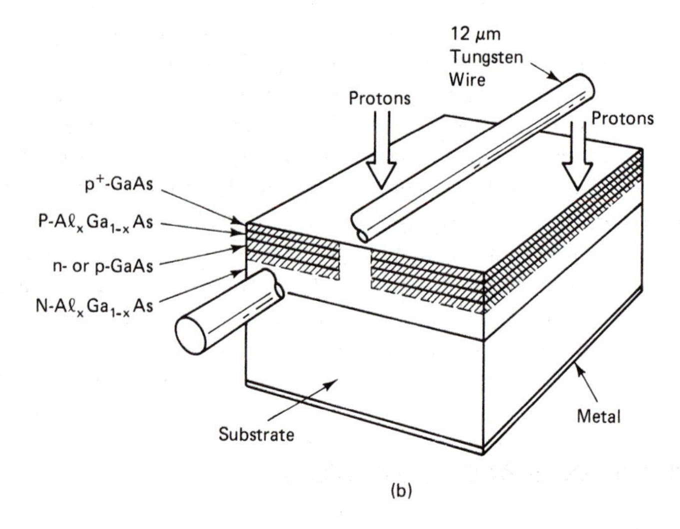

Figure 3-6 Double-heterostructure lasers (a), an oxide-defined stripe geometry laser diode, and (b) a proton-bombarded stripe-geometry laser diode. The tungsten wire is used as a mask to define the proton bombarded area.

spatial gain profile induced by carrier injection to provide lateral mode control. As the injected current increases, the gain is non-uniformly consumed by the stimulated recombination of carriers and the resulting gain profile deformation causes mode profile deformation or mode instability.[13,14] Laser diodes of this geometry are commonly referred to as gain-guided lasers and exhibit many longitudinal modes or spectral lines. A typical *L-I* and *I-V* curve, far-field beam divergence and mode spectrum for a gain-guided laser is shown in Figure 3-7a,b,c, respectively. Due to the changing gain profile, there tend

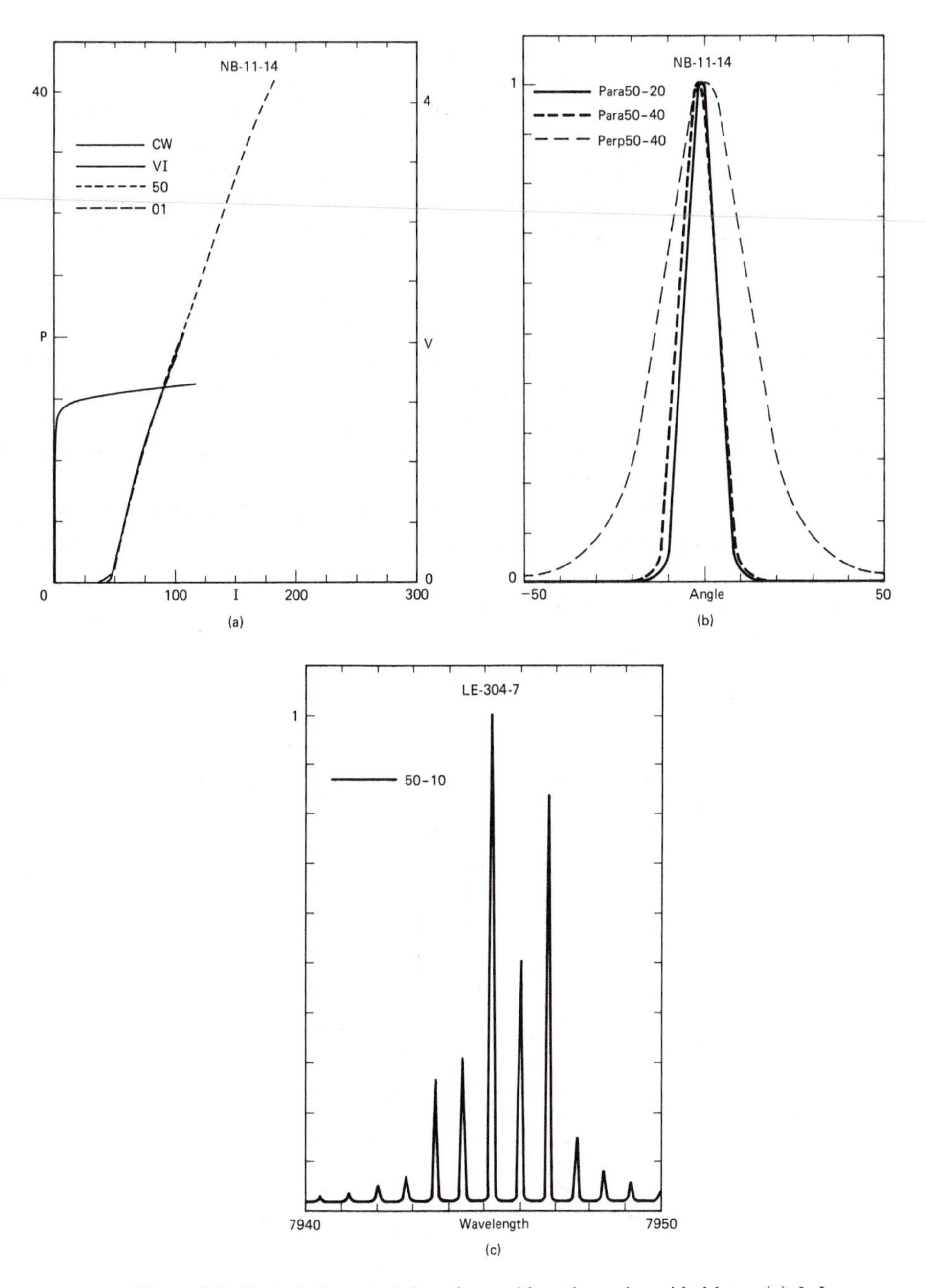

Figure 3-7 Typical characteristics of an oxide stripe gain-guided laser (a) *L-I* and *I-V* curves; (b) perpendicular and parallel beam divergences; (c) mode spectrum.

to be abrupt changes in the far-field pattern with corresponding "kinks" in the *L-I* characteristics as illustrated.

In addition to the far-field pattern being relatively unstable, gain-guided lasers are not useful in near diffraction limited systems because of large inherent astigmatism. Gain-guided lasers are gain-guided parallel to the junction, but index-guided perpendicular to the junction, which results in a curved wavefront measured parallel to the junction, and a flat wavefront measured perpendicular to the junction. The result is a difference in focal length of the beam of 20 μm or more when measured perpendicular and parallel to the junction. Astigmatism of even a few microns will significantly increase the divergence of the beam and decrease the intensity at the receiver as shown in Figure 3-8 from an analysis done by D. B. Carlin.[15] These effects can be corrected by cylindrical optics but the amount of astigmatism is in general not reproducible from diode to diode, and probably is not stable as a function of aging, temperature, or power output.

Astigmatism, kinks, and unstable multimode far-field patterns are not a problem when the lasers replace flashlamps for pumping solid-state crystals such as Nd:YAG. In this case, power efficiency and fairly narrow spectral outputs (±2 nm) to match the absorption peaks of the crystal are the important attributes and simple, reliable wide stripe (20 to 100 μm wide) gain-guided lasers are adequate.

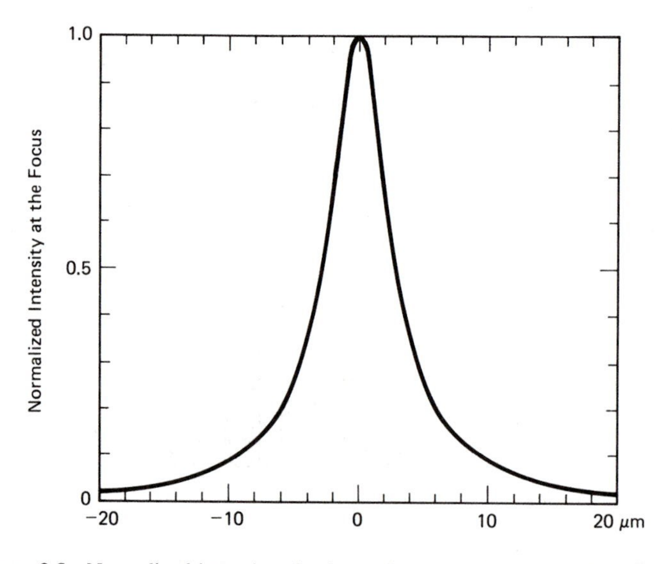

Figure 3-8 Normalized intensity of a focused laser diode beam as function of the difference in astigmatism measured perpendicular and parallel to the emitted wavefront.

3.3.2 Index-guided Lasers

An effective method of stabilizing the transverse mode and eliminating astigmatism is to introduce a built-in refractive index change along the junction plane. The buried heterostructure (BH) geometry achieves such an index change by embedding a high index-active region in a lower index material using a two-step growth,[16] as shown in Figure 3-9.

Propagation of only the fundamental mode in a dielectric waveguide requires that the product of the index change Δn between the "core" and the surrounding cladding and the square of wave-guide half-width, W, not exceed a fixed value. For the simple three layer waveguide shown in Figure 3-10, the relationship is

$$\Delta n W^2 = \frac{\lambda_o^2}{16(n_1 + n_2)} \tag{3-6}$$

where λ_o is the free space wavelength and n_1 and n_2 are the indices of refraction inside and outside the waveguide. However, for a BH structure, the resulting index change Δn along the junction plane is relatively large, about 0.01 to 0.1 so that stable fundamental mode lasing is possible only for waveguide widths of a few micrometers, resulting in low power and large beam divergences. In the early 1980s, more sophisticated structures were optimized to provide mode control with larger spot sizes; these include the channeled substrate planar (CSP) stripe,[17] the transverse-junction stripe (TJS)[18], and the constricted-double heterojunction (CDH),[19] all shown in Figure 3-11. Another structural consideration is the addition of an epilayer of intermediate index sandwiched between the active layer and an isolation (or cladding) layer. With proper design, the optical mode propagates mostly in the thicker waveguide layer—reducing the optical power density and the likelihood of

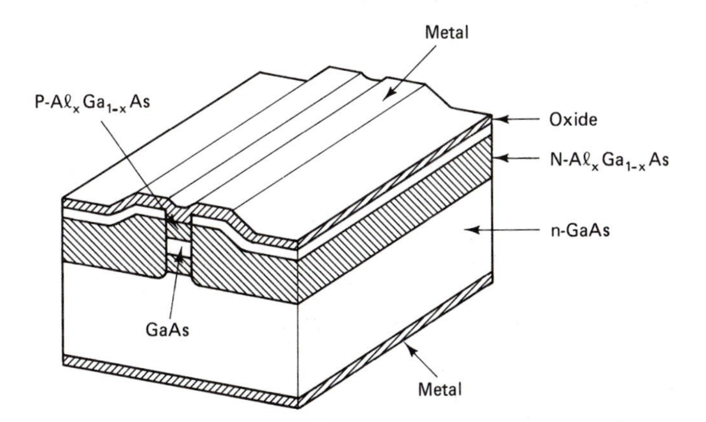

Figure 3-9 Buried heterostructure semiconductor laser geometry.

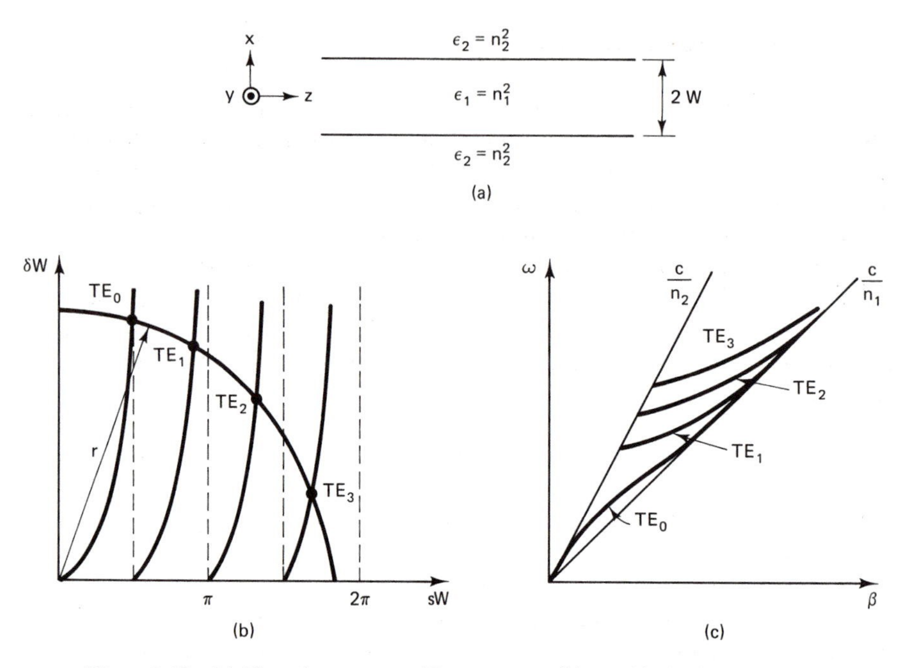

Figure 3-10 (a) Three-layer waveguide geometry; (b) graphical solution of the three-layer waveguide: sW and δW are the normalized transverse wave vectors in the center layer and the outer layers, respectively;

$$r = (n_1^2 - n_2^2)^{1/2} \frac{2\pi W}{\lambda_0}$$

(c) diagram for TE modes of a thin film waveguide.

facet mirror failure while obtaining electronic gain from the thin active layer. The resulting structure is referred to as a large optical cavity (LOC).[12] One trade-off with LOC devices is that the waveguide layer lowers the potential barrier against minority carrier injection and can result in a stronger dependence of laser threshold current with temperature than for the double heterojunction laser structure. It is found empirically for all semiconductor lasers that the threshold current has a temperature dependence given by

$$J_{th} \propto \exp(T/T_o) \tag{3-7}$$

For LOC devices, T_o ranges between 80 to 120° instead of the 140 to 190°C range typical of standard DH AlGaAs devices. InGaAsP lasers have even lower T_o's of 50 to 70°C which may ultimately limit their output power due to thermally induced saturation. An example of a LOC structure is the CDH-LOC geometry shown in Figure 3-12. Such LOC structures have exhibited some of the highest power outputs before facet damage.[4] Only a partial list of many existing semiconductor laser structures has been mentioned. However, all device geometries provide a perturbation in the plane parallel to the

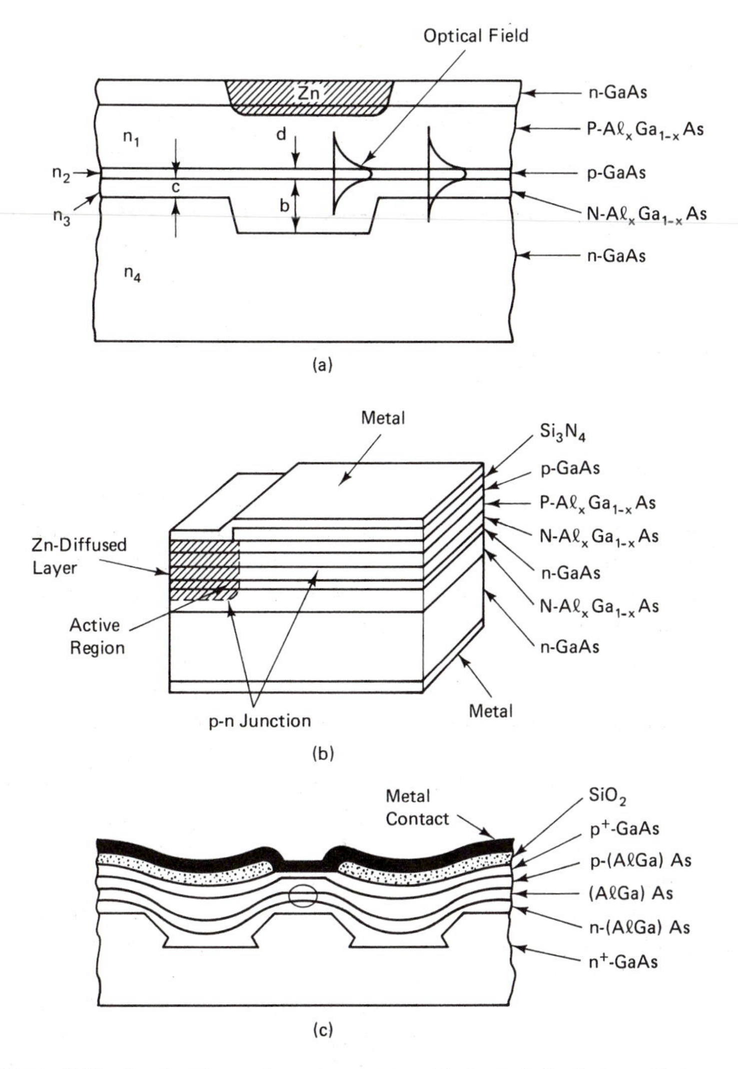

Figure 3-11 Semiconductor laser geometries: (a) channeled substrate planar (CSP), (b) transverse-junction stripe (TJS), and (c) constricted-double heterostructure (CDH).

junction to form a waveguide. One of the most popular geometries and easiest to fabricate is the CSP structure and its many variations.

CSP Device Design

Semiconductor lasers with geometries similar to the basic CSP design have obtained excellent results: power outputs up to 200-mW cw from a single device;[1,2] threshold currents of 30 to 40 mA; and total power efficiencies of >30 percent.

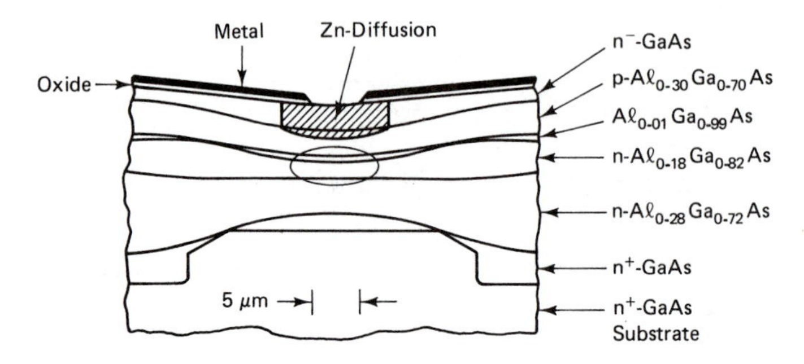

Figure 3-12 Current-confined constricted-double-heterojunction large-optical-cavity (CDH-LOC) semiconductor laser geometry.

To obtain a stable radiation pattern from a semiconductor laser during modulation requires that the built-in index change lateral to the *p-n* junction plane dominates over gain effects that change with injection current and spatial hole burning. However, as in the case of the BH geometry, if the index step is made excessively large, the lasing cross section must be kept small to maintain a single transverse mode. Due primarily to a shift in the absorption band edge, the index of refraction of the active layer changes with injected current until the laser reaches threshold. This phenomenon is referred to as "gain-induced index change"[20] and has the effect of reducing the index change lateral to the *p-n* junction plane by about 10^{-3}. Therefore, to obtain a stable radiation pattern, index-guided lasers need to have a built-in index change lateral to the *p-n* junction of greater than 10^{-3}—perhaps two or three times 10^{-3} which from Equation (3-6) limits the width of the spot size for AlGaAs lasers to a few microns (3 to 6 μm).

Figure 3-13 shows the amount of the real part of the effective index change lateral to the *p-n* junction for a CSP laser as a function of the thickness of the N-AlGaAs layer (Figure 3-11a) for active layer thicknesses of 400, 600, 800, and 1,000 Å. The indices of refraction are $n_1 = 3.41$, $n_2 = 3.62$, $n_3 = 3.41$, and $n_4 = 3.64$. From Figure 3-13, a CSP laser with a channel thickness b of 1.8 μm, and a "wing" thickness c of 0.3 μm has an unpumped lateral effective index change of $\sim 4 \times 10^{-3}$. Figure 3-14 shows the amount of the imaginary part of the index change lateral to the *p-n* junction for a CSP laser as a function of thickness of the N-AlGaAs layer for active layer thicknesses of 400, 600, 800, and 1,000 Å based on the same parameters used in Figure 3-13.

Figure 3-15 shows the calculated beam divergence perpendicular to the plane of the *p-n* junction for a CSP laser as a function of active layer thickness for the same indices of refraction for the CSP structure used in Figures 3-13 and 3-14. Also shown in Figure 3-15 is the relative peak-power density as a function of active layer thickness.

The beam divergence parallel to the plane of the *p-n* junction depends

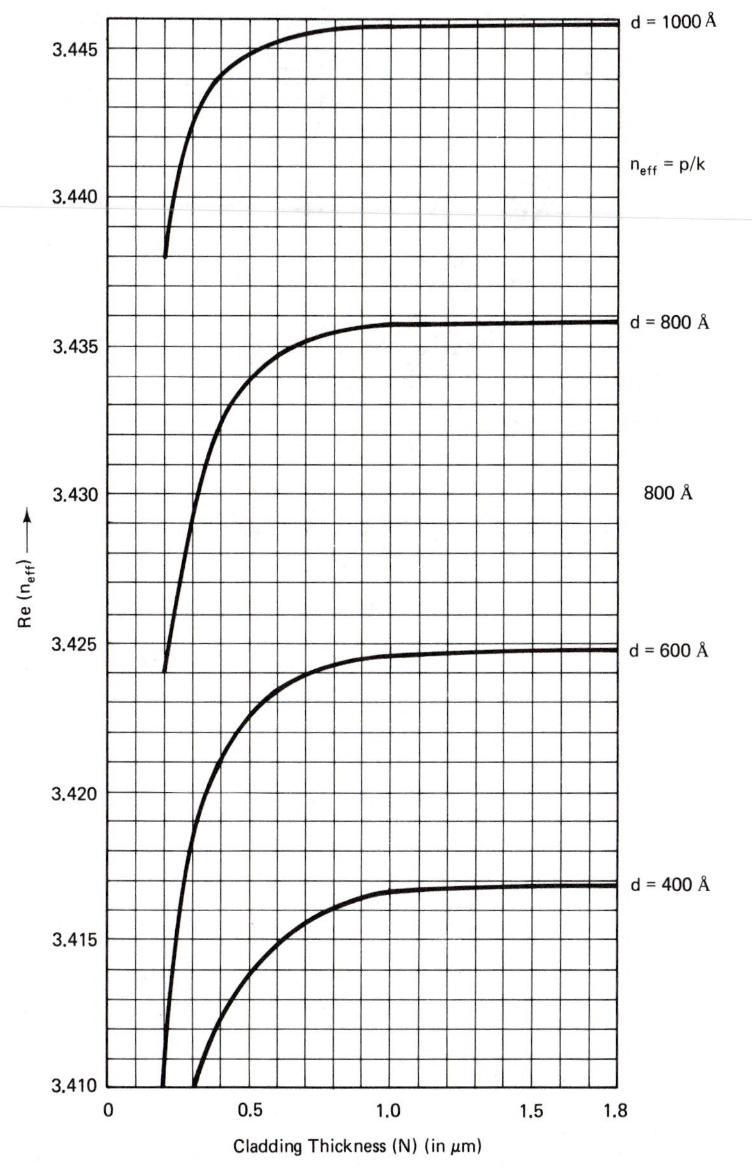

Figure 3-13 The real part of the effective index of a CSP structure as a function of the N-AlGaAs cladding layer thickness—dimensions *b* and *c* in Figure 3-11a.

on both the active layer thickness and the width of the CSP channel. For active layer thicknesses ranging from 400 to 1,000 Å, the perpendicular beam divergences can be 10 to 40° and for channel widths ranging from 6 μm down to 3 μm the parallel beam divergences range from about 6 to 12° at full-width half maximum (FWHM). The lower limits of beam divergence of ~16 × 5°

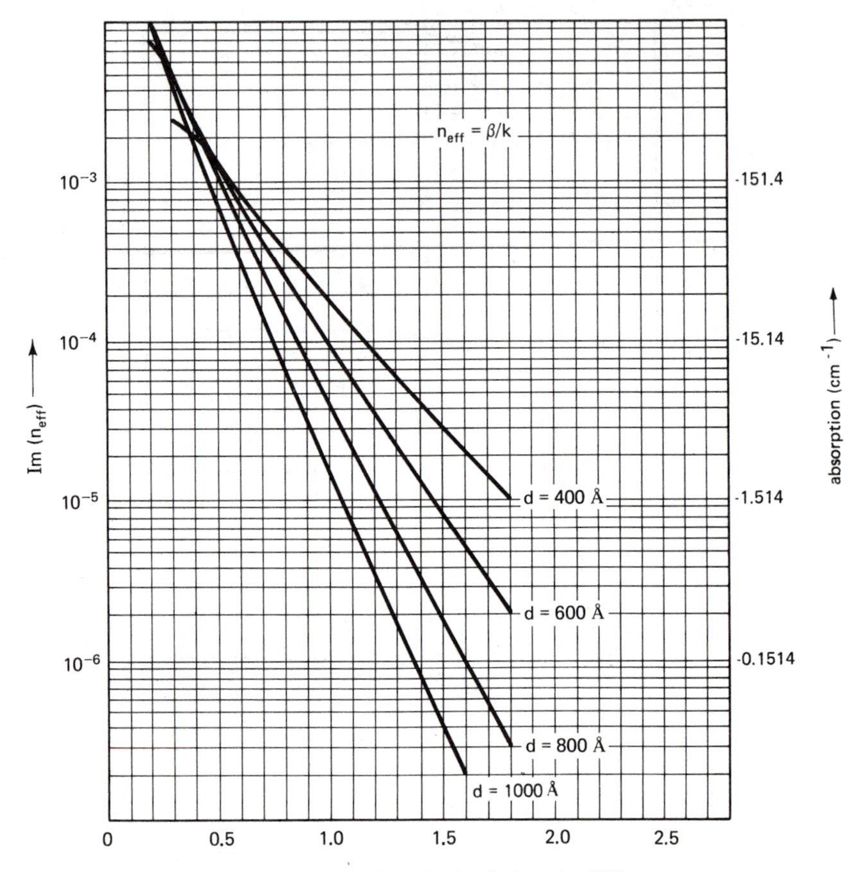

Figure 3-14 The imaginary part of the effective index of a CSP structure as a function of the N-AlGaAs cladding layer thickness. The power absorption ordinate is for a lasing wavelength of 0.83 μm.

are about the best that can be expected from any conventional single-emitter design, if stable operation during modulation or aging is to be maintained.

DFB and DBR LASERS

The single-mode spectral behavior illustrated in Figure 3-16 for a CDH laser is characteristic of all index-guided semiconductor lasers that employ cleaved facets as the laser mirrors. (Gain-guided lasers usually exhibit a number of longitudinal modes.) The transition from one mode to another is not a monotonic function of drive current, but can occur discontinuously over a 1 to 2 mA drive current change. Between hops, the mode shifts slightly with temperature, averaging 0.3 to 0.6 Å/K. Such mode movement and hopping is mainly due to increased active layer temperatures with increasing current levels. The injected carrier densities also change the energy-level separation for lasing, thus changing the wavelength. Also shown in Figure 3-16 are weak

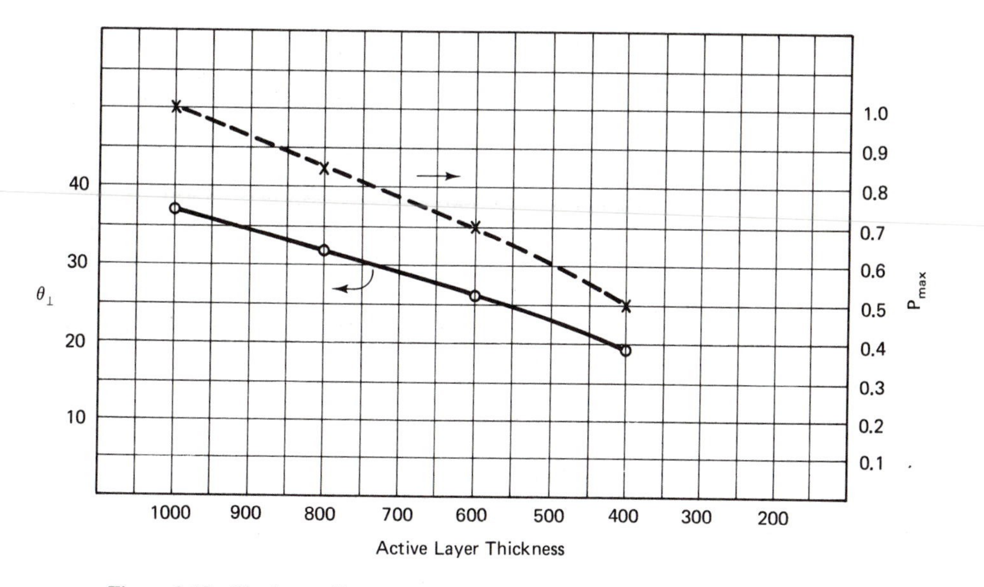

Figure 3-15 The beam divergence perpendicular to the junction and the relative peak optical power density as a function of active layer thickness.

satellite modes. A study on CSP lasers[21] has shown that the satellite modes may always be present at low power in some devices while in other devices the laser operates in these modes at full power a small faction of the time. Since it is an instability in the laser, mode hopping contributes significantly to the noise in optical communications systems. For stabilization against mode hopping, feedback from external[22,23] or internal[24-27] structures can be used. Such feedback provides the lasers with a multimirror cavity, which if adjusted properly, can only accommodate a single standing wave within the maximum gain region of the laser. Cleaved coupled cavity (C^3) lasers are a dynamically adjustable (via the current injection to one of the laser sections which alters the index of refraction) form of a multimirror approach to wavelength stabilization. Wavelength control has best been demonstrated using distributed feedback (DFB) and distributed Bragg reflectors (DBR). In this approach, the laser mirror is replaced with a grating either distributed throughout the laser or near the mirror facets as shown in Figure 3-17. Researchers at NEC have developed an InGaAsP DFB laser that has operated in excess of 50 mW.[27] Modulation (by direct modulation of the injection current) of a DBR laser shows that the lasing spectra remained single mode. Similar devices from the same wafer but without the grating, which were single mode during cw operation became multimode over 80 Å during rapid modulation.[26]

Very little work is presently under way on DFB and DBR AlGaAs semiconductor lasers, in part due to the technological difficulty in regrowing over a grating etched into AlGaAs. If the number of applications requiring a dynamic mode stabilized AlGaAs laser continue to remain small, perhaps the external third mirror,[23] external Bragg reflector (EBRL)[28] or cleaved coupled cavity[22] approach will be used. Such approaches trade off the complexity of semiconductor laser processing and design for external mechanical and electronic complexity.

In a crude sense, a DFB and DBR laser can be considered to be the extension of a third-mirror laser geometry to a multimirror laser geometry where the number of mirrors become several hundred to more than a thousand. If coherent heterodyne detection systems or very narrow band filters are to be used, spectrally mode-stabilized lasers such as DFB devices will be required. Even in direct detection, the lower noise inherent in mode-stabilized lasers improves system performance.

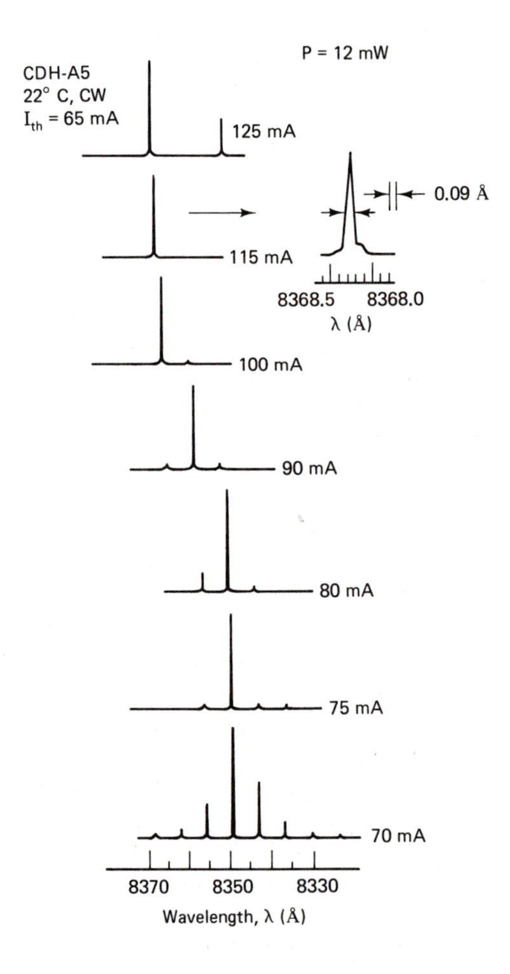

Figure 3-16 Spectral characteristics of a constricted-double heterostructure laser while driven to twice the threshold current value (P = 12 mW/facet).

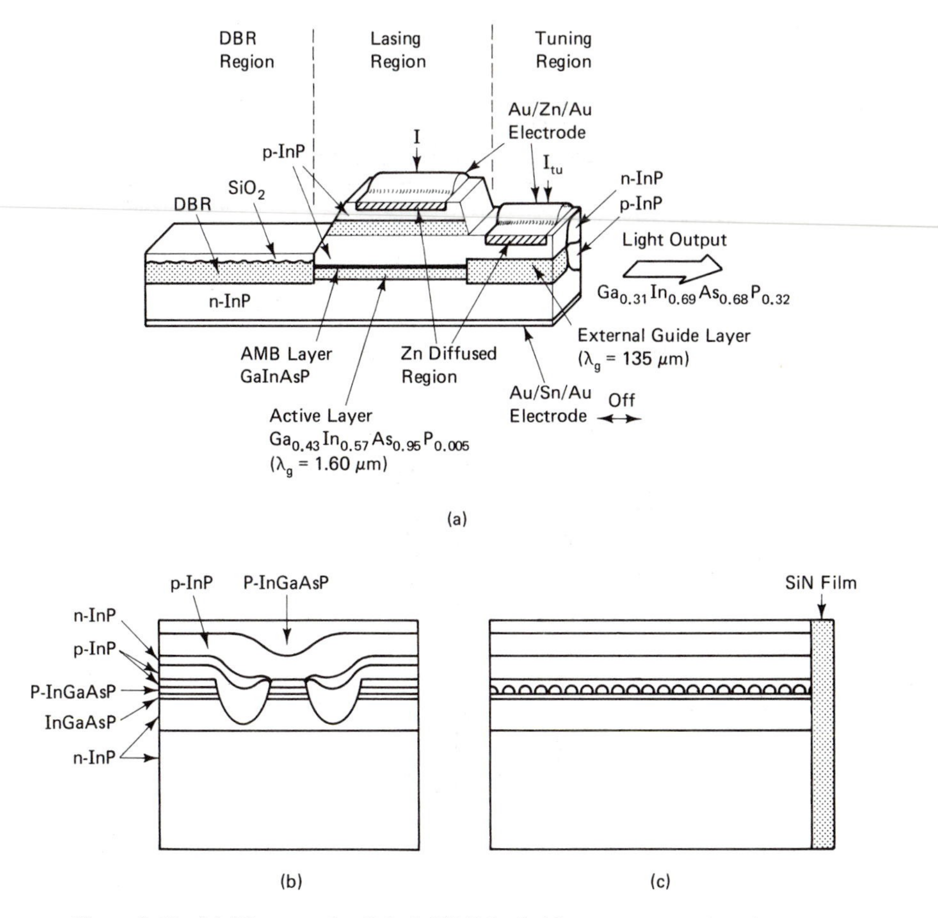

Figure 3-17 (a) Diagram of a GaInAsP/InP buried heterostructure, butt-jointed built-in distributed Bragg reflector (BH-BJB-DBR) integrated laser[25]; (b) cross-sectional view of a distributed feedback double-channel planar-buried heterostructure (DFB-DC-PBH);[27] (c) side view of a DFB-DC-PBH.

3.3.3 Phase-locked Arrays

Another class of semiconductor lasers that shows great promise as a source for space communications is phase-locked arrays. In this case, lasers are coherently combined on a single chip so that a single "phase-locked" beam is derived from several stripes. As shown in Figure 3-18, coherency for the simplest geometry is achieved by moving the stripes of gain-guided lasers within a few μm of each other and letting the optical fields overlap. Using this technique, 2.6-W cw has been obtained in a 40-element quantum well gain-guided stripe array with conversion efficiencies approaching 40 percent.[3]

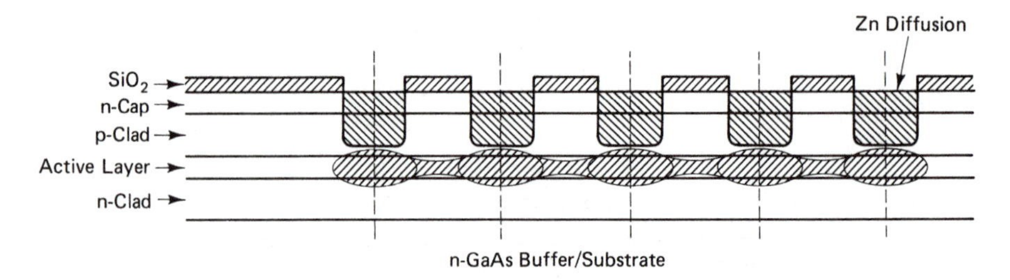

Figure 3-18 Geometry for a simple five-element gain-guided array.

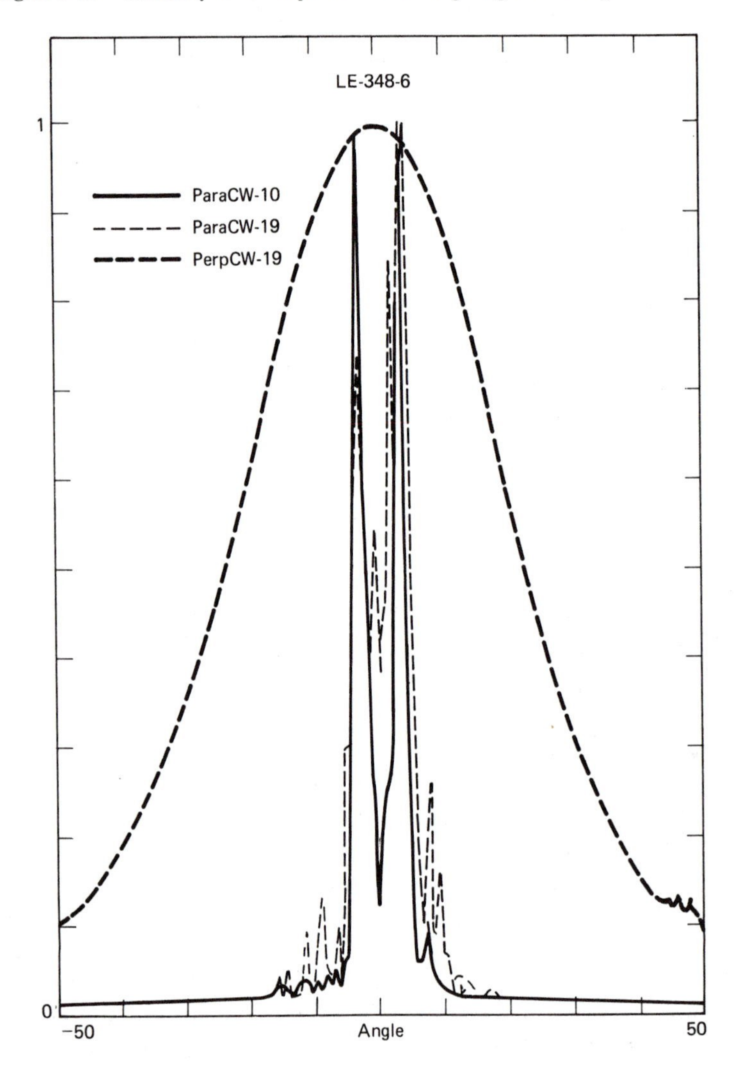

Figure 3-19 The dual-lobed far-field pattern corresponding to the structure shown in Figure 3-18. Note that the far-field pattern parallel to the junction is unstable with power.

However, such gain-guided phased arrays have the same problems as individual gain-guided lasers—i.e., unstable far-fields, multispectral modes, and astigmatism.

In addition, phase-locked arrays usually emit in a dual-lobed beam as shown in Figure 3-19. Because the regions between the array stripes are lossy, a simple-minded argument requires that the overall transverse electric field distribution of the array mode go through zero crossings between array elements to minimize optical loss. In this case, since the far field is related to the Fourier transform of the near-field array mode, the resulting far field will have two lobes. However, single-lobed far fields will be obtained if the near-field distribution can be shaped so that its Fourier transform is single lobed. Experiments show that providing gain between the array elements[29] or using a non-uniform element spacing[30] aid in shaping the far-field pattern of arrays. Since arrays have as many "low-order" modes (or supermodes) as stripes, the problem is even more complicated. Thus, it has been commonly found experimentally that most arrays have far fields that change their shape with power level (Figure 3-19). Figures 3-20 and 3-21 show the results of a theoretical calculation of far-field radiation patterns on a simple array of two CSP laser elements. By appropriate choice of geometry, the gain of the fundamental mode can be much higher than the next higher order mode (Figure 3-22), which is desirable for stability. Phased arrays of index-guided lasers are not as well developed as gain-guided arrays, but they hold the most

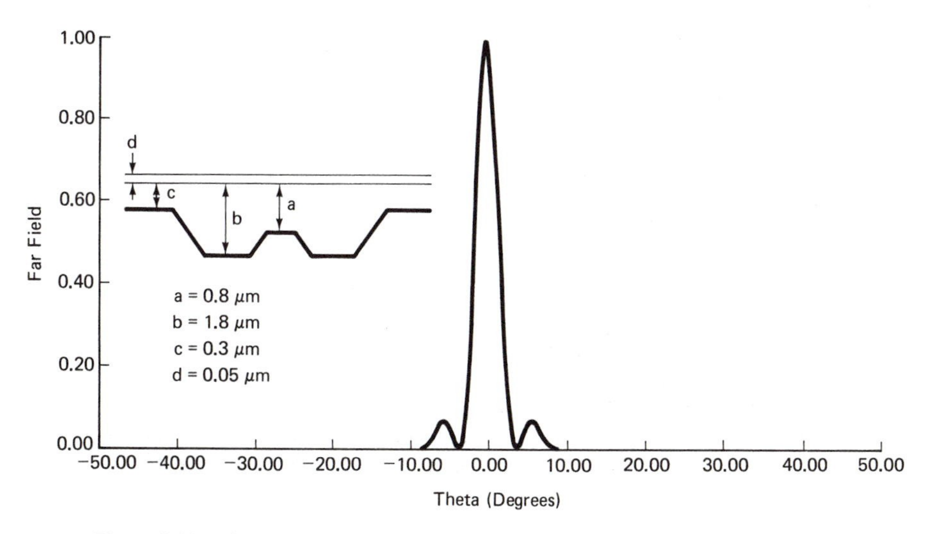

Figure 3-20 The calculated far-field pattern of the fundamental mode of a two-element CSP array with the geometry shown in the inset. The thickness *a* of the N-AlGaAs layer above the barrier is 0.8 μ, the channel depth is 1.8 μ, the thickness *c* of the N-AlGaAs layer outside the array elements is 0.3 μ, and the active layer thickness *d* is 0.05 μ. The channels are 5.5 microns wide at the top with a center to center spacing of 8.0 microns.

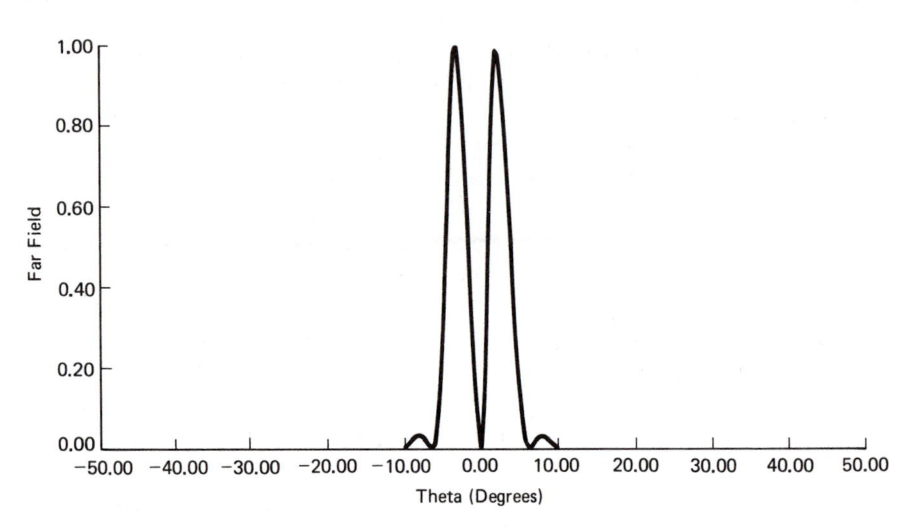

Figure 3-21 The calculated far-field pattern of the second mode of the two-element CSP array geometry shown in the inset of Figure 3-20.

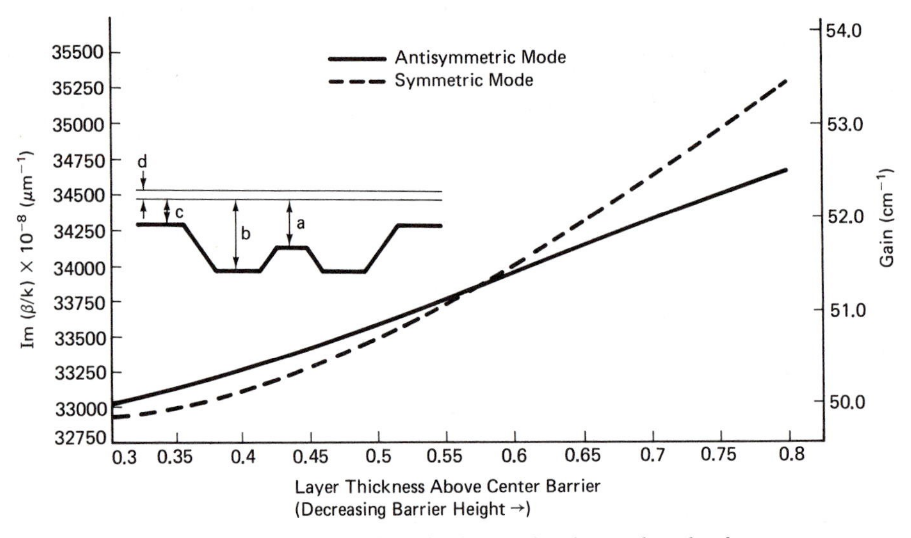

Figure 3-22 Calculated gain of the fundamental and second mode of a two-element CSP array as a function of the thickness *a* (see inset) of the N-AlGaAs layer above the barrier. The difference in gain between the fundamental and second mode increases as the dimension *c* increases, as the channels are narrowed, and as their separation is decreased.

promise for space communications applications. Power levels in these index-guided phased arrays are so far limited to about 80-mW cw in a single spectral mode and 200 mW in a single spatial mode,[31] but there appears to be no fundamental limit to the power output from arrays except for their stability against temperature and aging effects.

Since the beam divergences perpendicular and parallel to the plane of the junction of a semiconductor laser are related by a Fourier transform relationship to the perpendicular and parallel near fields, increasing the spot size in one plane decreases the beam divergence in that plane. For single devices, the spot size is narrowest perpendicular to the junction and widest parallel to the junction resulting in a larger beam divergence perpendicular to the junction (typically 16 to 35°). The small spot size perpendicular to the junction is a result of requiring a large bandgap difference (and hence large index difference) between the active layer and the confining layers to provide carrier confinement in the active layer in addition to optical confinement. Since only optical confinement is required parallel to the junction, the index step can be one or two orders of magnitude smaller parallel to the junction resulting in a larger spot size parallel to the junction. Typical beam divergences parallel to the junction range from 6 to 12°.

Most optical system designers require the perpendicular to parallel beam divergence aspect ratio to be 3 or less. Unfortunately, arrays spread out the spot size parallel to the junction resulting in even narrower beam divergences (as low as 1 to 2°). Using metal-organic chemical vapor deposition (MOCVD) or molecular beam epitaxy (MBE) material growth technologies, device geometries such as a quantum well graded-index (GRIN) structure (Figure 3-23) that uses separate mechanisms for carrier confinement and optical confinement perpendicular to the junction have been fabricated.[32] In addition to low threshold currents and high power efficiency, such devices can have perpendicular beam divergences as low as 6°.

Other concerns with arrays are yield and operating lifetime. The requirements of element-to-element uniformity for phased arrays and the effect of single-element degradation are not yet known. The only experience has been on gain-guided arrays where far-field or wavelength stability has not yet been established.

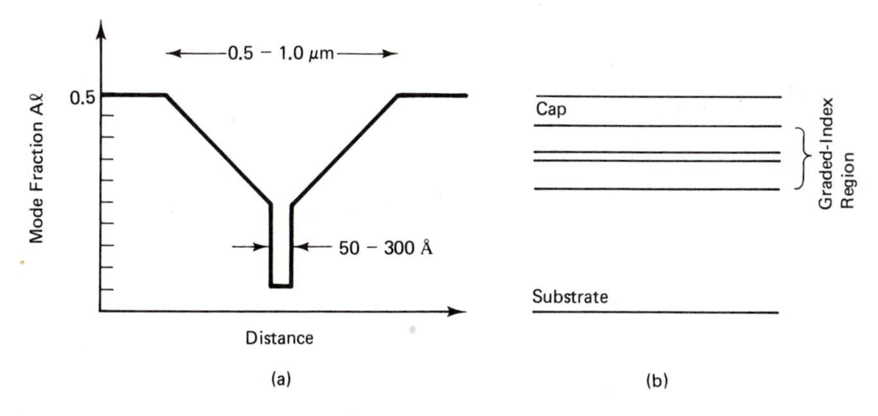

Figure 3-23 (a) Detailed drawing of the graded-index region indicated in the semiconductor laser device shown in (b).

3.4 OUTPUT WAVELENGTH CONTROL

3.4.1 Shorter Wavelength Operation

The lasing wavelength of AlGaAs laser diodes is determined by the mole fraction of AlAs in the active layer and today's AlGaAs lasers can be made to emit efficiently between 0.80 and 0.86 μm. However, as the lasing wavelength becomes shorter than 0.80 μm by increasing the mole fraction of AlAs in the active layer, room temperature cw operation becomes more difficult due to the following problems: (1) the threshold current significantly increases because of the smaller separation between the direct (Γ) and indirect (*L* and *X*) conduction-band minima in the active layer, and injection of carriers into the nonradiative indirect band; (2) a large compressive stress is imposed upon the active layer at room temperature because of the thermal expansion coefficient difference between the AlGaAs active layer and the GaAs substrate leading to poor reliability; and (3) it is difficult to obtain high conductivity in the Al-rich *p*-cladding layer leading to significant joule heating. CW lasing has been obtained at wavelengths as short as 6,800 Å[33] by replacing the GaAs substrate with a thick AlGaAs buffer layer above the DH structure to reduce stress and by using Mg as a *p*-type dopant instead of Ge or Zn. It is difficult to obtain high conductivity in Al-rich layers using Ge because of its reduced solubility and deep acceptor level. The use of Zn gives rise to cross-contamination problems in LPE growth due to its high vapor pressure. Mg has a large solubility, shallow acceptor level and low vapor pressure. However, so far, reliable high-power lasers have not been produced with wavelengths shorter than 0.78 μm.

Another technique that can be used to obtain shorter wavelength operation is the incorporation of quantum wells in the device. As the physical dimensions of the active region approach about 300 Å (the wavelength of an electron in GaAs), quantum effects become important. The density of states becomes quantized and the bandgap is effectively increased. Using a multiple quantum well (MQW) geometry consisting of seven 30-Å thick GaAs quantum wells separated by 50-Å thick barriers (with an AlAs mole fraction of 0.2), the lasing wavelength is reduced by approximately 1,000 Å.[34] Another important benefit of quantum well lasers is that the threshold current is significantly reduced.

By incorporating some of the above developments, reliable long-lived lasers with wavelengths well below 8,000 Å are anticipated.

3.4.2 Longer Wavelength Operation in AlGaAs Lasers

As the lasing wavelength increases above about 0.85 microns, the stress in the active layer changes from compression to tension for standard DH structures. The point of changeover varies with growth temperature, the bandgap energy difference between the cladding layer and the active layer, the mole fraction of aluminum in the layers, and the thicknesses of the layers.[35,36] A result of the calculation of such stresses is shown in Figure 3-24. For long life operation, it is important that the stress in the active layer not only have a low value, but also be compressive. Theoretically, this can be obtained with the geometry shown in the inset of Figure 3-24. However, the location of the stress minimum as a function of lasing wavelength can also be changed by varying the mole fraction of AlAs and the thickness of an AlGaAs buffer layer grown between the GaAs substrate and the first cladding layer.[36] Since a pure GaAs active layer emits at about 0.86 μm, compensating doping species such as Si, Te, or Zn can be added to the active region to form tail states which result in lasing at wavelengths as long as about 0.9 μm at room

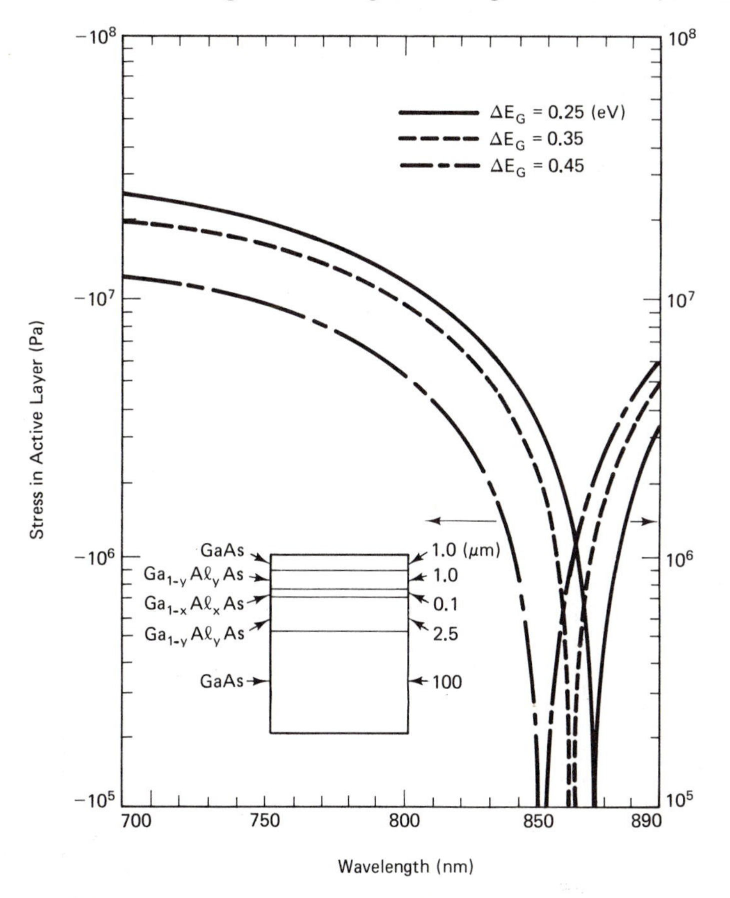

Figure 3-24 Variation of calculated stress in the active layer of a standard four-layered DH structure as a function of lasing wavelength. The bandgap energy difference between the cladding layer and the active layer is maintained at 0.35 eV. Layer thickness and AlAs mole fraction for the laser structure calculated in this figure are shown in the inset.[34]

temperature. The reliability of lasers emitting at 0.86 to 0.89 μm has not been adequately assessed.

3.5 SEMICONDUCTOR LASER LIFETIME

3.5.1 Background

Semiconductor laser operating lifetimes often are assumed to be sampled from a log normal distribution of failure time[37–39] similar to that reported for silicon and germanium diodes and transistors.[40] In a log normal distribution, as shown in Figure 3-25, the failure times on a logarithmic time scale have a Gaussian distribution. In Figure 3-26 are time to failure of oxide stripe gain-guided lasers at ~5 mW output at 70° and 23°C. To make such a plot more useful, such failure data (failure in this case being defined as the cessation of

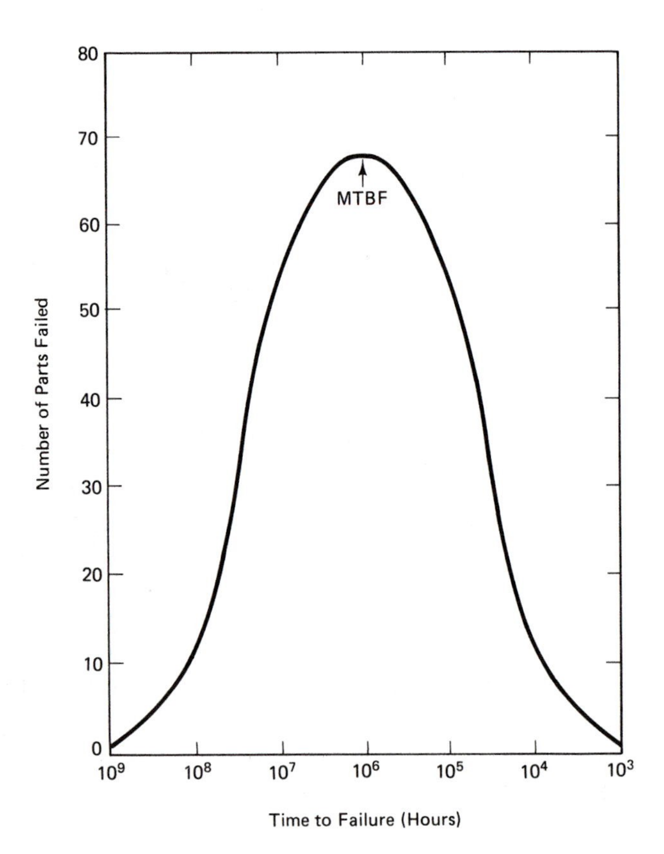

Figure 3-25 Number of failures as a function of time in a log normally distributed population showing the Gaussian nature of the distribution.

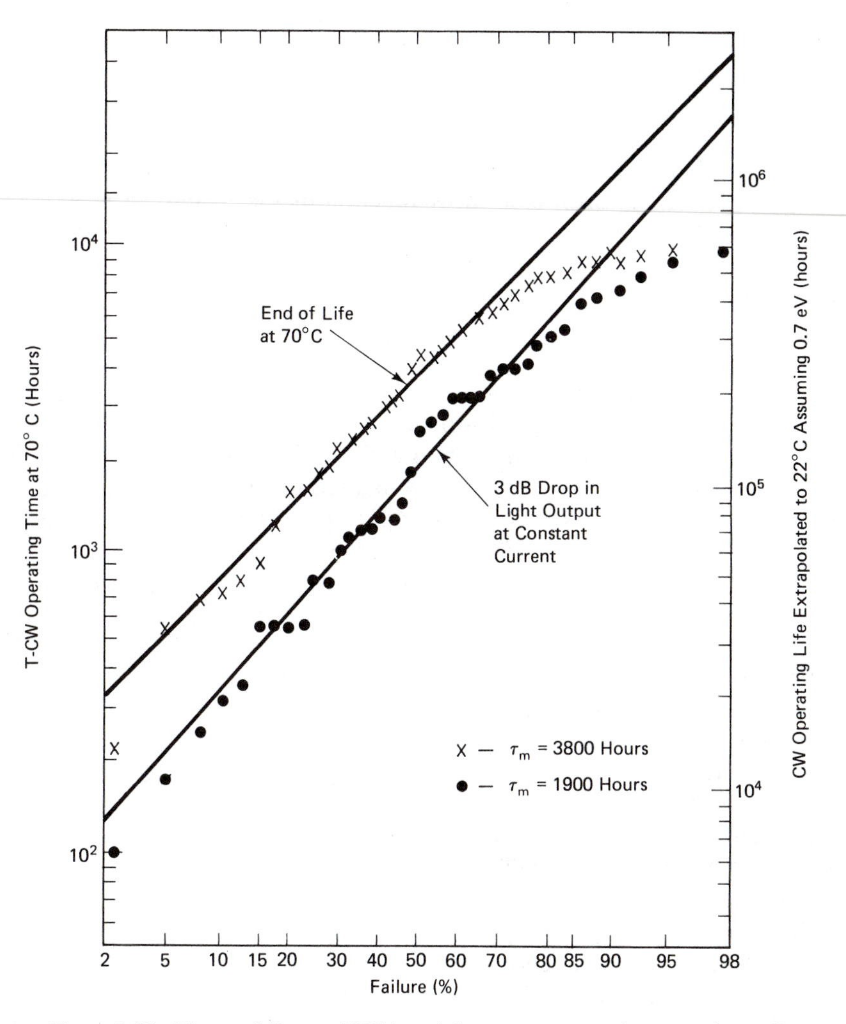

Figure 3-26 Time to failure at 70°C heatsink temperature on log normal coordinates for forty low-threshold (~50 mA) oxide-defined stripe lasers. Two failure points are assumed, the first when the laser drops to half its initial output at constant current (3-dB life) and the second when the laser can no longer emit 1.25 mW at the 70°C heatsink temperature (end of life). τ_m is the time needed for 50 percent of the lasers to fail.

cw output) is plotted on a log-normal axis such that the ordinate is logarithmic time and the abscissa a Gaussian of the percentage of the population failed.

For accelerated testing where temperature is the accelerating variable, the Arrhenius model is often used to extrapolate life data obtained from units tested at different constant elevated temperatures to obtain an estimate of the life distribution at usage temperature. The Arrhenius model is based on a theoretical relationship between chemical reaction rates and temperature.

The statistical assumptions behind the Arrhenius relation are (1) failure times at a given temperature are log normally distributed, (2) the log variance of failure times is constant over temperatures, and (3) there is only one failure mechanism with a corresponding activation energy. Mathematically, the mean $\mu(x)$ of the logarithmic life as a function of temperature is given by the Arrhenius relation, which is a linear function of the reciprocal $x = 1/T$ of the absolute temperature T; that is, $\mu(x) = \alpha + \beta$ where α and β are parameters of the material and the test method. The Arrhenius relationship at a given temperature T is defined as

$$\tau(t) = K \exp(E_a/kT) \qquad (3\text{-}8)$$

where $\tau(t)$ is the lifetime (the time for a specific degree of device degradation at a specified temperature T), k is Boltzmann's constant, K is a proportionality constant, and E_a is the activation energy. From the foregoing two equations, $E_a = k\beta$. The activation energy is a useful alternative quantity for expressing the dependence of life on temperature. Reported activation energies for oxide-defined stripe lasers range from 0.7 to 0.95 eV.[37,69,70]

Using the Arrhenius relationship for extrapolation of device data assumes a single failure mechanism over the temperature range. At low power outputs, the main failure mechanism is internal damage or degradation that is the formation of nonradiative defects within the active region of the laser. These defects are formed during forward-bias operation and are caused by the energy released during (nonradiative) electron-hole recombination, which moves or creates nonradiative centers. The specific nature of these defects has not been absolutely identified but their formation is a function of current density and temperature. At high power outputs other failure mechanisms may become important, as will be discussed.

In Figure 3-26, Ettenberg and Kressel[8] have plotted the fraction of failed oxide-defined stripe-geometry devices during operation at approximately 4-mW cw at 70°C as a function of time.

On the left-hand ordinate, the operating time at 70°C is plotted and on the right-hand side the estimated operating time at 22°C is indicated assuming 0.7 eV activation energy in the Arrhenius relation.

For the same population, two arbitrary failure points are noted. The first is the time at which the laser output falls to half of its initial value at constant current; the second is the time at which the laser, after repeated increases in current to maintain light output, can no longer emit more than 1.25 mW at 70°C. The two failure points chosen are useful for laser systems employing a temperature-compensated drive, thermoelectric cooling, or optical feedback to maintain constant output. If we ignore the deviation from a single log normal population in Figure 3-26 (a straight line on this type of grid) that occurs after about 75 percent of the devices failed (indicating either a non-log-normal population or more than one log normal population), 50 percent of the devices can be predicted to fail in a few hundred thousand hours at room temperature based on device performance at 70°C.

The deviation from log normal statistics exhibited in Figure 3-26 is a common occurrence in semiconductor laser life tests, indicating a need for more effort directed at understanding the statistical failure distributions and complex failure mechanisms of these devices. Additionally, informal reports from two laboratories have shown instances in which lasers under life test lived longer at higher temperature (70°C) than at lower temperature (50°C). As long as there is a mixture of failure mechanisms in lifetests, only engineering estimates rather than probabilistic estimates of laser diode lifetime and reliability can be made. Such deviations from well-behaved log-normal statistics probably represent in general the developmental nature of the devices tested to date and the need for process control to improve the uniformity of the devices.

Other investigators[41-44] have performed accelerated lifetests based on device failure at high current (I) levels, on the assumption that device lifetime was proportional to I^{-n} where n has values between 1.5 and 2.0. Accelerated lifetests at high current levels are commonly used for quaternary lasers because they characteristically have low T_o's and therefore are difficult to operate at high temperatures. A composite Arrhenius-type relationship that includes both temperature and current factors has been empirically derived and compared with experimental data[45].

A present problem of matching accelerated life tests with actual time tests at operating current and temperatures is that the nonaccelerated life tested data are always outdated because the technology for material growth, processing, fabrication, and the device geometry is continuously changing. Laser life-testing technology is still at the stage at which some life testing from each individual wafer is considered necessary before lasers from that wafer can be deemed useful for long life.

For space and other high reliability applications, the selection of devices by screening[46] or purging[47,48] are current areas of investigation.

3.5.2 Semiconductor Laser Aging Mechanisms and Characteristics

Introduction

During operation and before failure (however defined), a combination of changes in the laser operating characteristics can occur. Shifts in spectra and far-field patterns, along with self-sustained oscillations, occur and may not be accompanied by decreased power (at constant current) or an increase in the threshold current.

The causes of laser diode degradation can be separated into three areas: facet (mirror) damage; ohmic contact degradation; and internal failure, which includes dark line defect (DLD) formation. Facet damage and contact degradation become more important as the power level of the device is increased.

Catastrophic Mirror Damage and Mirror Erosion

Attempts to operate semiconductor lasers at high power can result in gross damage to cleaved facets that act as mirrors[49].

The critical peak optical power density for catastrophic mirror damage is approximately 10^6 W/cm^2 for cw operation[50] and about 10^7 W/cm^2 for pulsed operation with very short pulses (<100 ns)[51].

The power level at which catastrophic damage occurs in a given laser has been found to vary with the reciprocal of the square root of the pulse length for pulse lengths under about 1 μs[52,53]. For pulse widths substantially longer than 1 μs, the mirror damage is the same as would result from cw operation.

The limit on optical power density at the facet of AlGaAs semiconductor lasers is believed to be due to a high surface state recombination rate which prevents the facet surface from being inverted, thus causing absorption of laser light. This can either melt the mirror surface in a catastrophic manner or propagate dark lines and other defects into the active region. Interestingly, the longer wavelength InGaAsP semiconductor lasers (with corresponding lower energy photons) do not seem to have a limit on the optical power density at the facet. This may be due to a lower recombination rate at the surface of InGaAsP. For AlGaAs lasers considerable work has been done to eliminate light absorption at the facet by "window" or nonabsorbing mirror (NAM) geometries. By increasing the band gap within about 10 to 25 μm of

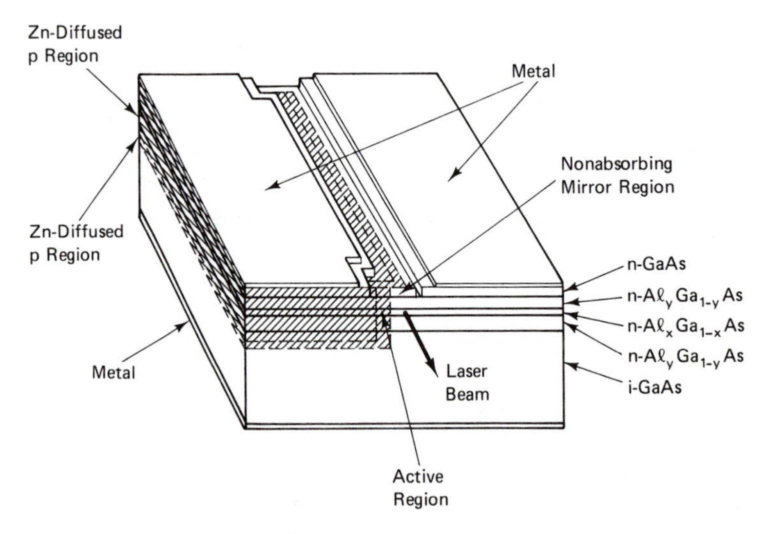

Figure 3-27 Diagram of the TJS laser with nonabsorbing mirror (NAM) structure.[54]

the laser facets (mirrors), the lasing photons are not absorbed and the facet damage limit is increased by about a factor of 10. The influx of dark lines and defects are also reduced leading to longer operating lives even at power levels which do not lead to mirror damage.[54] Several methods have been used to achieve a nonabsorbing facet. One structure uses a zinc diffusion only over the inner region of the stripe, which depresses the bandgap except near the facets. This structure, known as the crank TJS and shown in Figure 3-27, is commercially available and rated at 30 mW. The conventional TJS is only rated at 10 mW. Since the TJS is mounted with the active region away from the heat sink, thermal resistance is high (about 100° C/W) and the power is limited for the TJS by thermal rather than facet damage considerations. A second approach completely removes and replaces the active region near the facet with a high Al content AlGaAs layer during an additional growth step. This approach has been used on both buried heterostructure[55] and CDH-LOC[56] geometries (Figure 3-28). Such NAM-CDH-LOCs have been able to

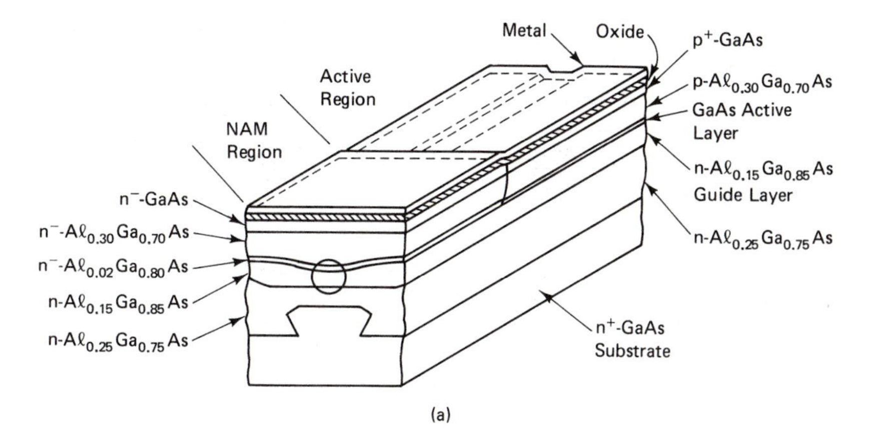

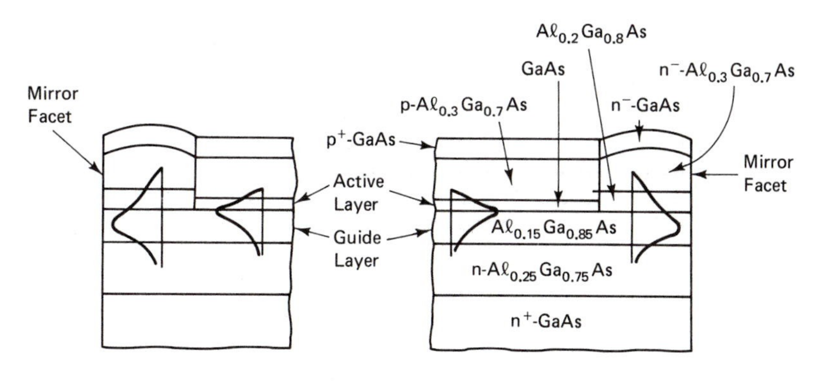

Figure 3-28 (a) Nonabsorbing-mirror (NAM) CDH-LOC structure; (b) longitudinal cross-section of NAM-CDH-LOC laser.

achieve pulsed power outputs in excess of 1 W before mirror damage. A third approach uses a difference in channel width at the facet to give a thinner active layer near the facet which results in less absorption.[57] This structure is called a V-groove substrate inner stripe (VSIS) and is shown in Figure 3-29.

In AlGaAs lasers, facet erosion occurs at power levels well below the catastrophic damage limit and results from oxidation of the mirror facet at the active region by a process activated by emitted radiation in the presence of water vapor.[52,58] Facet erosion decreases the mirror reflectivity and increases the nonradiative recombination rate at the facets which then lowers the internal efficiency and increases the threshold current. Protection from facet erosion is given by facet coating as shown in Figure 3-30.[9] The facets were coated with a film of Al_2O_3 which inhibits oxidation to preserve facet reflectivity, and hence the threshold current. The facet coating inhibits oxidation of the facet which occurs during lasing operation; the oxidation is accelerated in moist ambients. Facet coatings having an optical thickness of

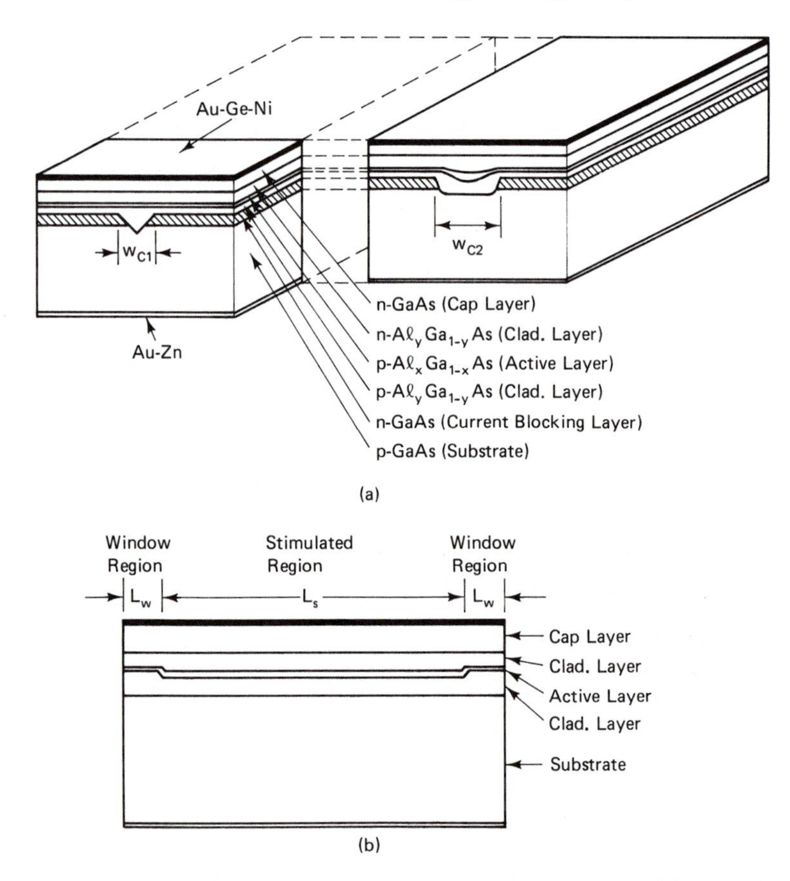

Figure 3-29 (a) Schematic representation of a window VSIS laser; (b) cross-sectional view at the center of the channel of a window VSIS laser.[55]

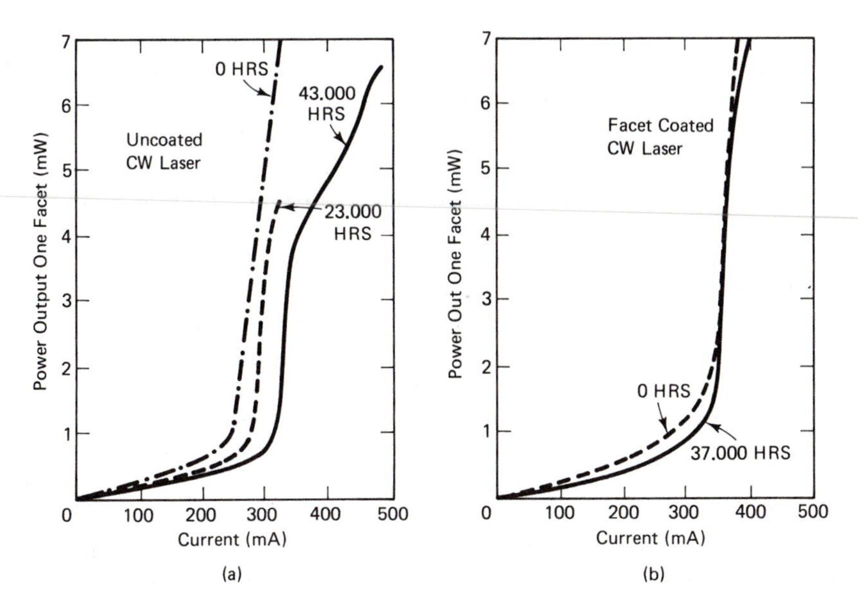

Figure 3-30 Cw power versus current characteristics at various times during the room temperature cw life tests of an (a) uncoated laser operated in a dry ambient at ~4 mW output and (b) an Al_2O_3 coated laser operated in a laboratory ambient at 10 mW output.

one half wavelength leave the facet reflectivity and therefore the threshold current unchanged. Since the optical damage limit depends on the peak internal optical power density, facet coatings can be used to reduce the peak internal optical power density for a fixed output power. The internal power P_{in} is related to the output power P_{out} by

$$P_{in} = P_{out} \times \frac{1 + R_f}{1 - R_f} \tag{3-9}$$

where R_f is the facet reflectivity. A common practice in fabricating high power lasers is to use a quarter wavelength antireflection coating on the front facet, and a high-reflectivity coating on the rear facet. The actual facet reflectivities fabricated are about 5 to 10 percent for the low and 80 to 90 percent for the high-reflectivity coating. The threshold current remains approximately the same because it depends only on the product of the two-facet reflectivities. However, the internal power is reduced by about one half for a fixed output power, allowing greater margin before the onset of catastrophic optical damage. Simultaneously, the front facet power efficiency is about doubled.

Internal Damage

Internal damage in the laser, which increases the threshold current and sometimes decreases the differential quantum efficiency, often appears as dark line defects (DLD). A DLD is a network of dislocations that can form during laser operating in the region of the active cavity.[59] Once started, it can grow extensively in a few hours. The initiation of DLDs can take much longer. The DLD produces relatively high absorption and is a region with a high density of nonradiative recombination centers. The growth of DLDs is accompanied by an increase in the threshold current and a decrease in the external differential quantum efficiency. Studies are not in agreement on whether the DLDs originate at a point outside the active layer[60] or are confined to the active layer.[61,62]

DLD growth can be extremely erratic. If the DLD forms at all, it is usually observed within hours after the start of cw operation. However, in some cases, the start of DLD has been delayed for hundreds of hours. The DLD growth rate has been observed to vary greatly without apparent reason and sometimes to cease abruptly after a period of rapid growth.

Recent work with quaternary (InGaAsP) DH lasers indicates the DLD formation is appreciably slower than with AlGaAs lasers.[63] A possible explanation is that the narrower energy-gap material provides less nonradiative recombination energy for defect motion than does AlGaAs[64] or the difference in atom[65] sizes of the InGaAsP act to pin the growth of the dislocations via stress differentials.

Elimination of DLD is believed possible by the prevention of the source dislocations by careful bonding of the laser to the heatsinks to minimize strain. Thus, if the dislocation density of the substrate is relatively low (less than $1{,}000/cm^3$), each 250 x 10 μm stripe-geometry laser has a small (less than 4%) chance of incorporating a dislocation if no new defects are introduced during crystal growth or device fabrication. A burn-in test period can be used to eliminate most devices that fail rapidly because of DLDs.

During LPE growth, particulate matter generates stacking faults and dislocations in GaAs-AlGaAs layers,[66] and oxygen contamination[67] can also have a profound effect on defect concentration. The use of small amounts of aluminum in the active region is believed to improve reliability because of the gettering effect of aluminum on oxygen.

The success that has been achieved in reducing DLD formation in fabricated lasers has resulted from the care paid to cleanliness and purity during heteroepitaxial wafer growth, and the use of fabrication techniques that eliminate bonding strain. Careful inspection for defects in the grown wafer has also been beneficial.[68]

Ohmic Contact Degradation

The deterioration of ohmic contacts is a problem common to all semiconductor devices and is observed under certain conditions in transistors, rectifiers, and other components subjected to either high current densities or high temperatures. Because the threshold current of laser diodes is fairly temperature dependent, an increase in thermal and/or electrical resistance affects the laser's performance. In laser diodes, the thermal resistance of the contact between the laser chip and the heatsink tends to increase with time.[69–71] This degradation process depends on the solder used, the current density through the contacts, and the temperature. Until about 1982, most laser diodes were soldered with indium, because indium melted at a low temperature and provided a soft, flexible bond to the heatsink that did not stress the diode. A possible mechanism for the degradation of indium-soldered lasers is the formation of intermetallic compounds involving indium and gold, which have a relatively high thermal resistivity.[8] Indium migration may also occur, as suggested by the observation of voids in the solder after long-term operation.[72]

The increase in the thermal resistance of the contact produces an increase in the junction temperature for a given operating current. Thus, because the cw threshold current is increased, it is possible to observe a decrease in the cw output of a laser operated at constant current that is directly caused by this effect. One technique for differentiating between degradation resulting from deteriorating contacts and that resulting from other causes is to monitor the low duty cycle pulsed-operation threshold current and compare it to the cw laser threshold. In Figure 3-31, an example is shown of a laser that degraded solely because of thermal resistance increase, as evidenced by the unchanged pulsed-operation threshold current and substantially increased cw threshold after 21,000 h of operation.[8] The electrical resistance of the laser may also increase with time, but this effect usually is not as severe in its impact as the thermal resistance increase. A possible explanation of the non-log-normal distribution in Figure 3-26 is an increase in thermal resistance because of contact resistance that accelerates the degradation as a result of internal damage.[8]

A reduction in the rate of increase in drive current (to maintain a constant output power), threshold current, and thermal resistance of DH lasers has been reported with harder and higher melting temperature solders, such as gold-germanium or gold-tin used with diamond, silicon, or BeO submounts. The results of one study[71] are summarized in Table 3-1. Laser operation for over 1,800 h (70°C, gold-tin eutectic solder) and over 10,000 h (70°C, gold-germanium eutectic solder) have been reported with a slight increase in driving current.

Especially for low-power applications, mounting the lasers with the junction side up shows promise. In effect, the substrate acts as the submount. Overall degradation rates of below 10^{-4} per hour for AlGaAs lasers at 100°C have been reported[73] in this configuration. Additional processing techniques such as laser alloyed,[74] pulsed-electron-beam annealed[75] contacts, gold-to-gold bonding, or new contact materials may further improve ohmic contact performance.

As might be imagined, the very high-power and high-efficiency opera-

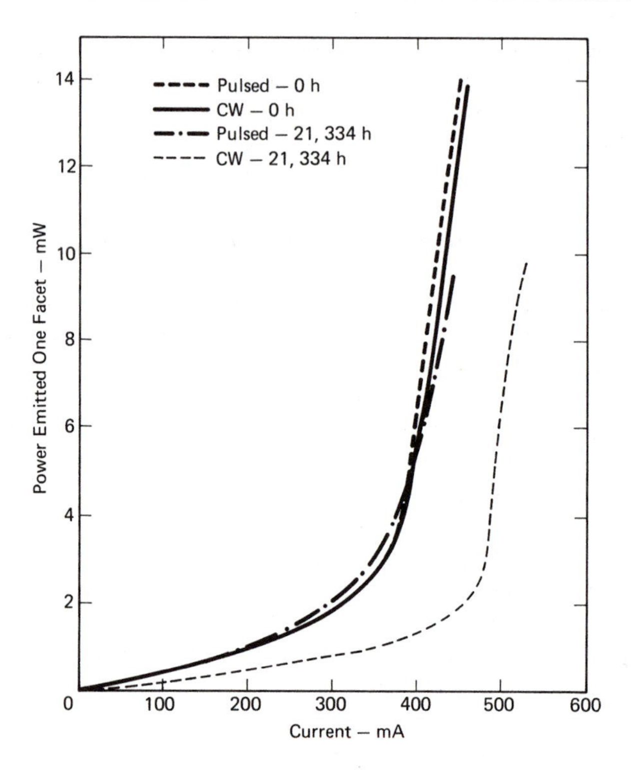

Figure 3-31 Power output as a function of pulsed (0.1 percent duty cycle) and direct current operation of a laser after operating 21334-h cw at room temperature with a very high current level of 450 mA. Whereas the pulsed and dc thresholds were very nearly the same initially, a substantial difference is seen after long-term operation because of the diode's thermal resistance increase. As a result, the power emitted in constant-current operation is reduced.

TABLE 3-1 COMPARISON OF AGING CHARACTERISTICS AT 70°C WITH RESPECT TO THE LASER BONDED WITH EACH SOLDER.[71]

Measured Parameter	*AuSn*	*AuGe*	*In*
a_r of I_d(%/kh)[a]	1.0	1.0	6.6
a_r of I_{thc}(%/kh)[a]	0.4	0.6	3.6
a_r of R_t(%/kh)[a]	~0	~0	150
Stress	Observed ~10^9 (dyn/cm^2) at room temperature	Observed ~10^9 (dyn/cm^2) at room temperature	Not observed

a_r = rate of increase

tion required for space communication places an increased premium on low thermal and electrical resistance and maintaining these low values over the required system operating time.

3.5.3 Degradation Effects on System Performance

A laser may fail to perform satisfactorily in a system long before it ceases to lase. Previously, increases in threshold current and drive current (to maintain a given power level) that occur with aging were discussed. Two other significant changes in laser characteristics may occur: changes in the mode spectra (lasing wavelength, far-field pattern, spectral width) and self-sustained oscillations in the laser output.

The normalized light output at constant current for a typical long-life laser that was aged at 70°C is shown in Figure 3-32. Also shown is a slight shift of several angstroms to longer wavelengths with aging. After many thousands of hours of continuous operation, with the operating current typically increased to maintain the power output, the spectrum is broadened and shifted to longer wavelengths by as much as 100 Å.[6] In some instances as shown for index-guided CDH lasers, the spectrum can also shift to shorter wavelengths, while still maintaining single spectral spatial outputs. In a system that uses laser diodes to pump a Nd:YAG laser, the laser diode output could

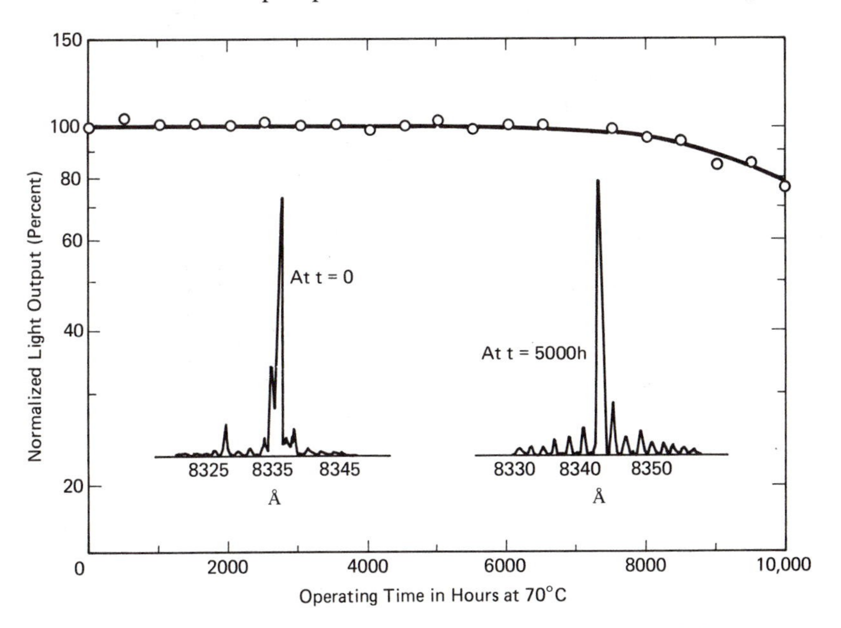

Figure 3-32 Typical spectra shifts with aging. The constant-current light output versus time for the same long-lived DH laser operated at 70°C ambient is also shown.[6]

shift from an absorption peak to an absorption minimum (Figure 3-33). In Figures 3-34 and 3-35, changes in the lateral far-field with aging are shown[8,37] for gain-guided lasers. These mode changes are basically alterations in the beam direction and shape. Laser applications that rely on fixed-mode properties, such as coupling into optical fibers or conventional optics will be limited by such degradation. There is some indication that severe far-field and wavelength changes may occur only near the end of laser life.[8,37] The spectra in Figure 3-35, however, correspond to the same laser emitting 5 mW at room temperature in both cases.

In single-mode index-guided lasers, spectral hops are instabilities that introduce noise into the signal. Such noise increases in turn require higher than anticipated signal (power output) levels from the transmitter. An increased number of such mode hops for a given modulation can occur during aging.

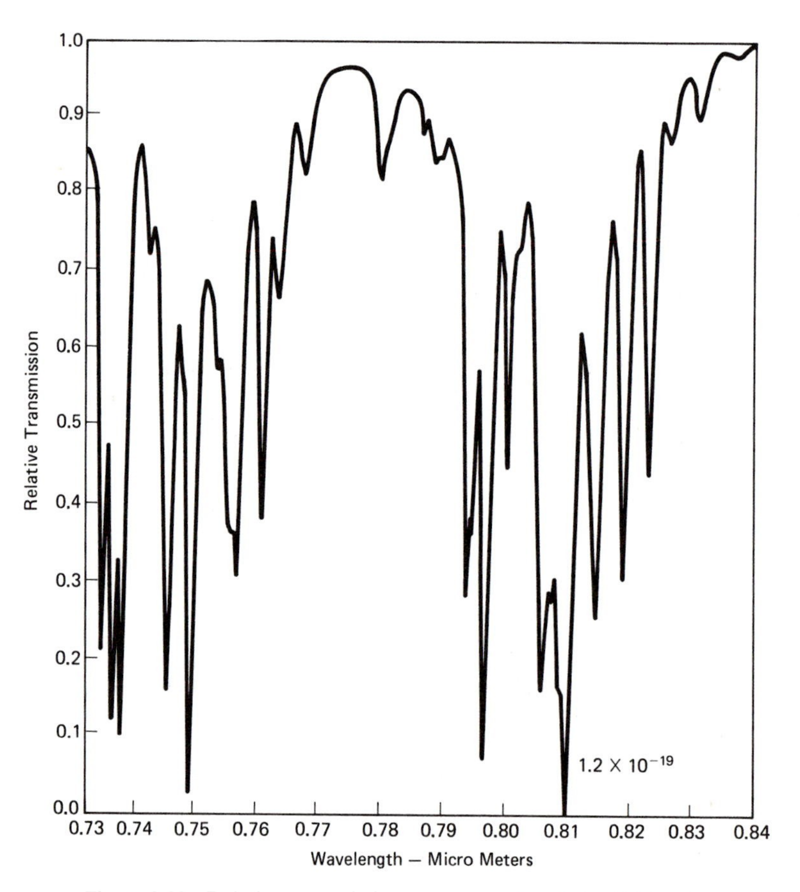

Figure 3-33 Relative transmission versus wavelength for Nd:YAG.

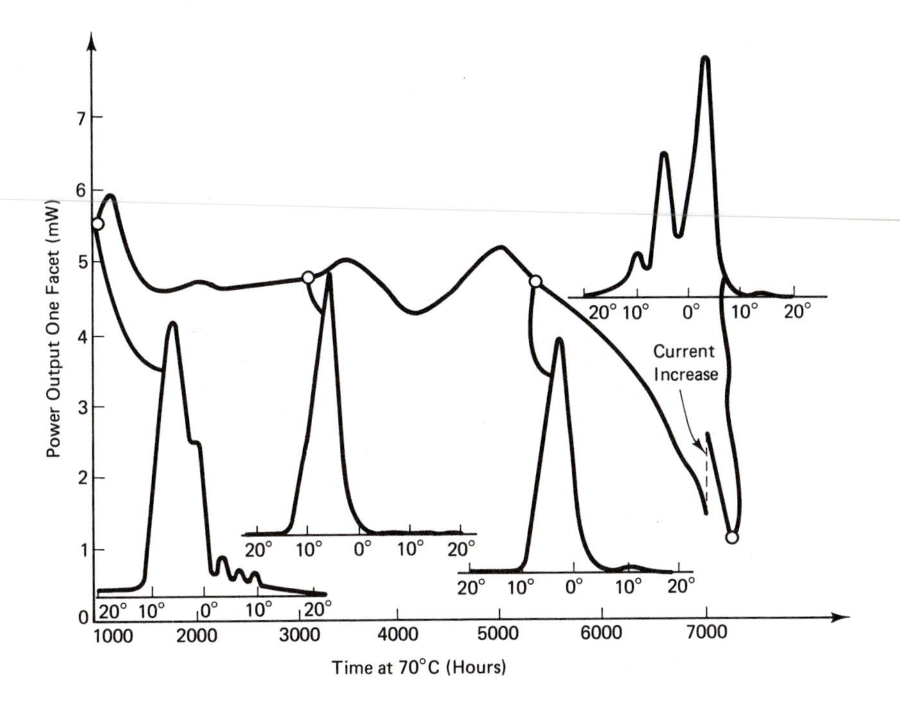

Figure 3-34 Light output as a function of time at 70°C heatsink temperature of oxide-defined stripe laser showing changes in the lateral far-field during accelerated life tests.[8]

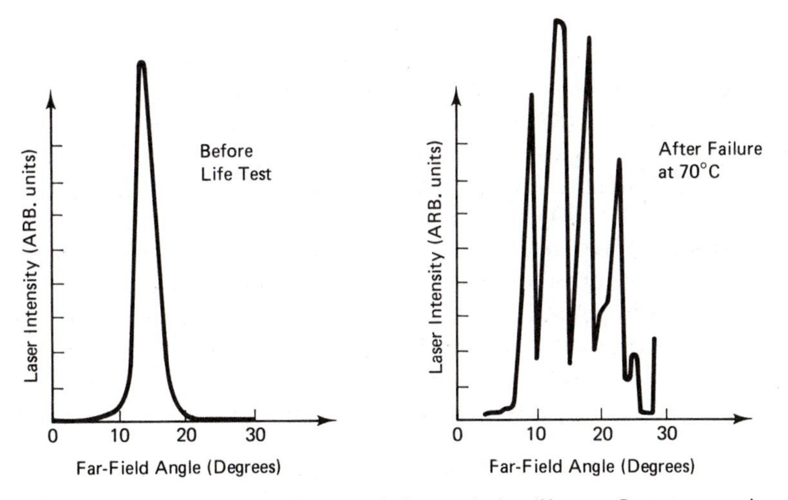

Figure 3-35 Typical far-field pattern before and after life test. Output power is 5 mW in both cases.

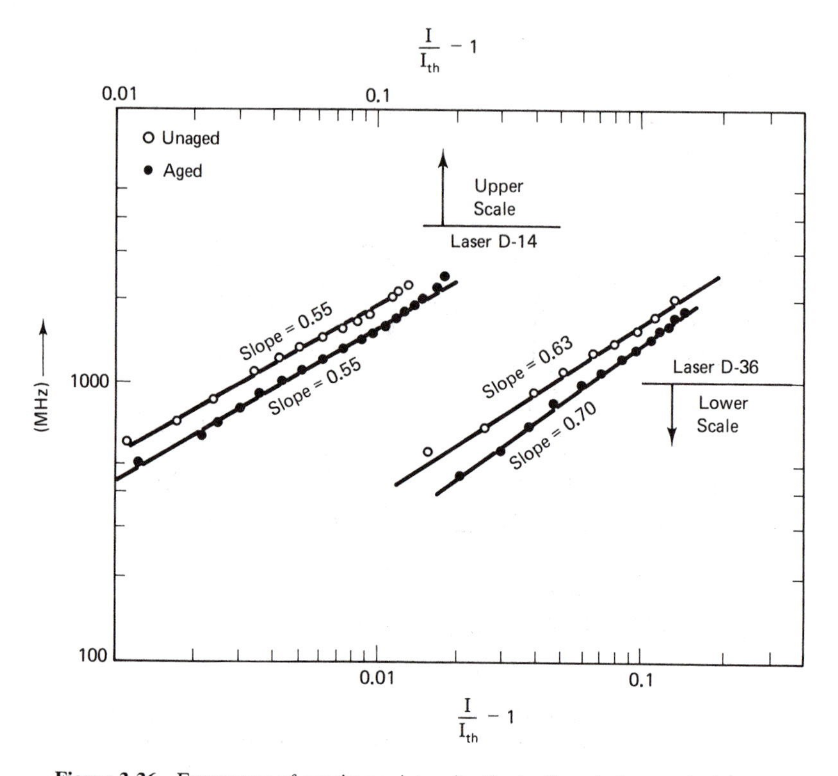

Figure 3-36 Frequency of maximum intensity fluctuations before and after aging as a function of level above threshold for the cw double-heterostructure lasers that developed weak intensity modulation during continuous operation at 70°C for 50 h (Ref 76).

Another characteristic of diode lasers that is relevant to operation in optical communication systems is the tendency for the diodes to develop self-sustained oscillations (SSO) of their optical intensity. These pulsations can occur at the beginning of operation or after aging. Frequencies of the SSOs range from 150 MHz to about 2 GHz;[76,77] these pulsations are illustrated in Figures 3-36 and 3-37. Figure 3-36 illustrates the dependence of the frequency of the SSOs on drive current above threshold.[76] In earlier work, SSOs were reported to be uncorrelated to changes in threshold current density, axial nonuniformities in pumping current, or DLDs. However, later work related SSOs to DLDs (in the vicinity of the facets) and to mirror erosion.[58,78] For the suppression of SSOs, SiO_2 coatings on the facets were reported more effective than Al_2O_3.[78] SSOs have also been suppressed by placing an external mirror close to a laser facet.[79]

A final consideration, in applications that take advantage of the ease of directly modulating the light output of semiconductor lasers, is the reduc-

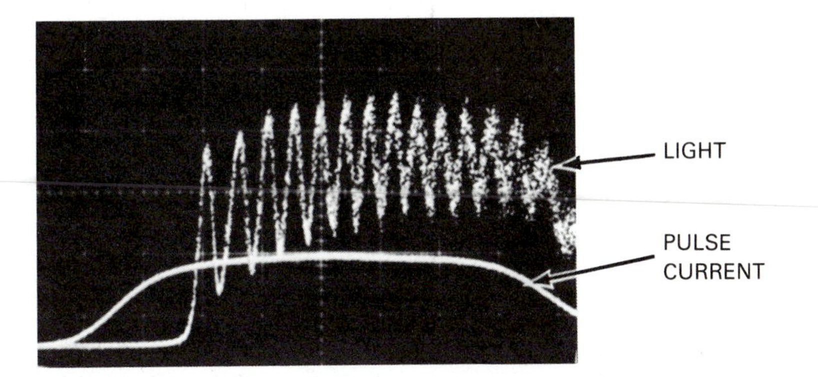

Figure 3-37 Time variation of the injected current pulse and the output intensity from a DH laser exhibiting self-sustained oscillations.

tion in depth of modulation caused by changes in the location of the knee of the light output versus current curves. Dixon and Dean[80] reported a fourfold reduction in system lifetime attributable to this effect in an optical transmitter using proton bombarded stripe lasers.

3.6 MODULATION

Optical communications with injection lasers, except for schemes that employ semiconductor lasers as pumps for a solid-state laser such as Nd:YAG, involves modulating the optical output intensity by applying a corresponding modulation to the laser drive current. Considerable research has been devoted to the response of laser diodes to isolated pulses, digital data streams, and sinusoidal modulation, and to the optical noise and self-modulation produced by the laser independently of external modulation. These phenomena have generally been approached individually on a limited variety of device geometries. On this basis, there is a growing understanding of their nature. There is lacking, however, complete data on all these interrelated phenomena as they appear in typical and well-characterized examples of the various device types.

Current understanding of laser modulation is based on consideration of the coupled dynamics of the carrier concentrations ($N(x,t)$) and the photon concentration ($P(x,t)$), within the active region of the device. Since the active region is very small (on the order of 0.1 μm or less) compared to the carrier diffusion length and the optical mode size, the lateral position x (parallel to the junction plane) is the only spatial variable required and t is the time dependence. These coupled rate equations have been written in many equivalent forms and can be generalized to include multiple spatial and longitudinal modes[81]:

$$\frac{\partial N(x,t)}{\partial t} = \frac{J(x,t)}{ea} - \frac{N}{\tau_c} + D\frac{\partial^2 N}{\partial x^2} - \sum_{ij} G_{ij}(N)|E_j(x,t)|^2 S_{ij}(t) \tag{3-10}$$

$$\frac{dS_{ij}(t)}{dt} = \left[G_{ij}(N) - \left(\frac{1}{\tau_p}\right)_{ij} \right] S_{ij} + \frac{\gamma_{ij}}{\tau_c} \int N(x,t)dx \tag{3-11}$$

$$S(x,t) = \sum_{ij} S_{ij}(t)|E_j(x,t)|^2 \tag{3-12}$$

In the equation, i and j refer to longitudinal and transverse modes. The active region has thickness a and is composed of material with carrier lifetime τ_c and carrier diffusion constant D. Photon lifetime $(\tau_p)_{ij}$, spontaneous emission-coupling factor γ_{ij}, and stimulated optical gain $G_{ij}(N)$ are specified for each mode. J represents the current density of injected carriers that pumps the carrier population inversion.

In general, the waveguiding properties of the laser depend on the complex refractive indices of each layer and the gain profile, which are in turn influenced by G_{ij}, N, and J. The gain is a function of both the carrier concentration N, and the photon concentration S. A complete analysis starts by solving the modal electric field distribution $E_j(x,t)$ for the semiconductor waveguide geometry with no injected current. Then by an iterative process, the effects of carrier diffusion, current spreading, spatial hole burning, and thermal effects may be included.

Although numerical techniques are often required, considerable information has been obtained from a small signal analysis of the rate equations,[82-84] which results in a modulation response with the transfer characteristics of a second-order low-pass network.[12] A resonance in the modulation response, known as a relaxation oscillation in lasers, results from the interplay between the optical field and the population inversion. The relaxation oscillation frequency obtained from the small signal analysis is[84]

$$f_\tau = \frac{1}{2\pi}\sqrt{\frac{AP_0}{\tau_p}} \tag{3-13}$$

where

A = differential optical gain
P_o = steady-state photon density
τ_p = photon lifetime, given by

$$\tau_p = \frac{1}{v[\alpha + 1/L\ (\ln 1/R)]} \tag{3-14}$$

v is the group velocity of the light, α is the distributed loss, L is the length of the cavity, and R is the mirror reflectivity. The small signal modulation bandwidth of a semiconductor laser is usually considered to be equal to f_T.

Equation (3-13) relates the relaxation frequency to three fundamental independent parameters: the differential optical gain constant A, the photon lifetime τ_p, and the photon density P_o. Thus, by increasing either the optical gain coefficient or the photon density, or by decreasing the photon lifetime, the small signal modulation bandwidth can be enhanced.

The modulation responses of a 120-μm long BH laser at various bias levels are shown in Figure 3-38.

Increased small signal modulation rates have been demonstrated by operating the laser at low temperature,[84] which increases the optical gain coefficient. Both quantum well structures and detuned cavity loading have been proposed[84] as a means to increase the optical gain coefficient at room temperatures.

Raising the small signal modulation rate by operating the laser at high power is only limited by catastrophic optical damage or thermal considerations.

The photon lifetime can be decreased (therefore, increasing the small signal modulation bandwidth) by shortening the length of the laser cavity.

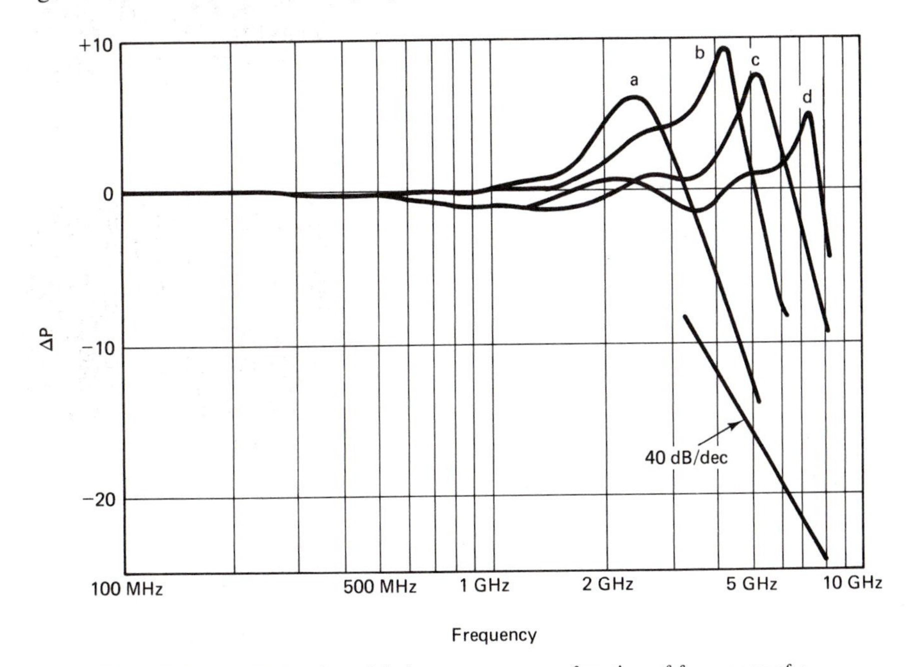

Figure 3-38 Small signal modulation response as a function of frequency of a 120-μm BH laser on a semi-insulating substrate at the following bias levels: (a) 1 mW, (b) 2 mW, (c) 2.7 mW, and (d) 5 mW. The modulation power ΔP is normalized at low frequency.[84]

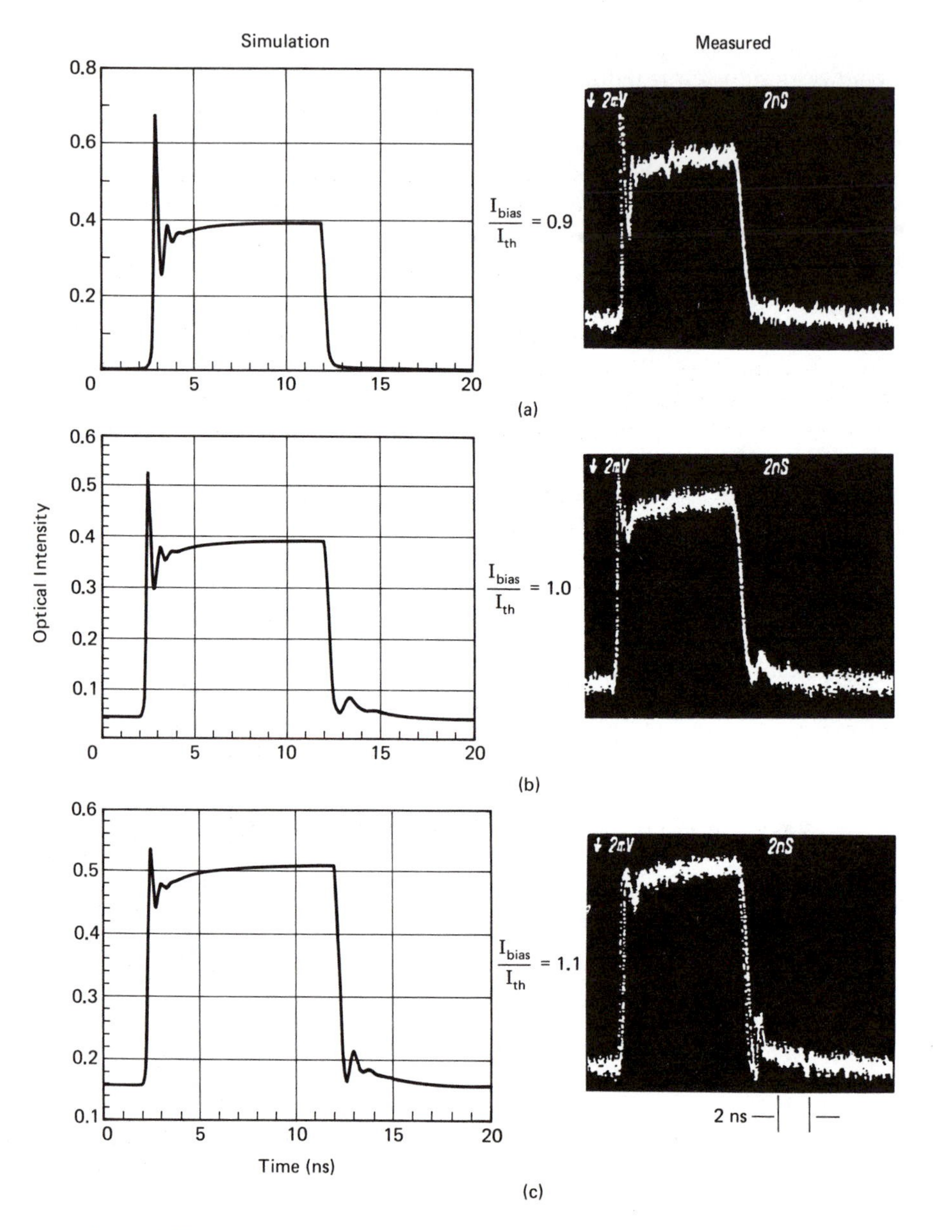

Figure 3-39 Analytical and experimental measurements of a CSP laser diode under pulse modulation at (a) $I_{\text{bias}}/I_{the} = 0.9$, (b) $I_{\text{bias}}/I_{the} = 1.0$, and (c) $I_{\text{bias}}/I_{the} = 1.1$.

However, this approach is limited in that shorter lasers (100 to 200 μm or less) require excessively high current densities for a given power output.

By optimizing these parameters, small signal modulation rates in excess of 12 GHz at room temperature and 23 GHz at $-70°$ C[9,10] have been obtained.

For space communication over large distances, the on-off ratio in the laser output should be as large as possible to increase the signal to noise ratio. The modulation response of a laser under large modulation is generally worse than under small signal modulation, and can be quite different. However, theoretical analysis of deep modulation of semiconductor lasers[85] has given excellent agreement with experimental results and predict that very high-power lasers (approximately 100 mW) can be digitally modulated in excess of 1 GHz. Figure 3-39 shows the computed and experimentally measured modulated waveform for a high-power CSP laser diode under pulse conditions.

3.7 RADIATION EFFECTS

Comparatively little work has been done on the effects of space and nuclear radiation environments on semiconductor lasers. Work has been reported mainly on laser diodes of the oxide-defined-stripe and proton-bombarded-stripe geometries. The available data indicate that diode lasers are rather hard to radiation levels of 10^{14} neutrons/cm^2, 10^7–10^8 rad total dose of γ radiation, 10^{11} rad/s X-rays, and such doses do not cause catastrophic failure. Instead, irradiation appears to speed up aging characteristics—e.g., increases in laser threshold current, lasing wavelength shifts, mode structure changes, and increased turn-on time delays.

Increases in laser threshold current have been reported as a result of electron, gamma,[86,87] and neutron[88–90] irradiation caused by a decrease in radiative efficiency η_i. Similar changes in laser threshold occurred with proton bombardment,[91] but the proton induced defects acted primarily as hole traps rather than as recombination centers. Consequently, in contrast with other types of irradiation, proton bombardment increases the optical absorption α but does not significantly affect the radiation efficiency η_i. The introduction of additional nonradiative traps by neutron irradiation has been suggested as an explanation for increases in the laser turn-on time delay.[92]

Changes in the wavelength and mode structure of the laser output spectrum following electron and gamma ray irradiation have also been observed.[86]

3.8 FUTURE EXPECTATIONS

For space communications, the single most important requirement for a laser diode is cw output power. Increasing laser output power decreases satellite weight, cost and complexity dramatically. Fortunately, in the last several years, dramatic increases in the performance of semiconductor lasers have been achieved. Transverse stabilized single spatial and single longitudinal mode AlGaAs lasers have exceeded 100-mW cw in the laboratory and cw output power as high as 200 mW from a single AlGaAs laser has been achieved. Conceivably, these high-power laboratory "champion" devices will become commercial products rated at 100 to 200 mW. With the perfection of non-absorbing mirrors (NAMs), 250 mW may be obtainable in a single device. The limitations on the peak internal optical power density while simultaneously maintaining a stable, single spatial mode and a single lobed far-field presently result in difficulty in surpassing 100 mW in AlGaAs laser diodes and will limit the optical output power of single devices to well below 1 W for the foreseeable future. However, all these problems are solvable and advances are rapidly being made in material quality, device processing, and device design.

Dynamically stabilized diode lasers will be commercially available at the longer wavelengths (1.3 to 1.55 μm) in the late 1980s. If substantial commercial and government markets for optical memory and optical space communications develop, dynamically stabilized short wavelength AlGaAs lasers will be developed also. The results of dynamic stabilization of DCPBH lasers by DFB[27] show that no penalty in power results from using an internal grating.

Since high optical power densities are consistent with high modulation rates, modulation rates for commercial high-power lasers approaching the 10 GHz range will be achieved.

Present laser diodes are expensive. Such costs are not inherent. With production of laser diodes increasing for fiber optic and audio disc applications, costs will decrease when the production exceeds the present 10^3–10^4 lasers/month for most manufacturers and reaches 10^5–10^6 lasers/month, as it will in the next five years. Laser diodes will then be sold for a few dollars apiece. At quantities greater than this level, laser diodes must and will sell for less than a dollar.

Current research may lead to additional commercial products in the next two to ten years: cw lasers emitting in the ranges from 0.6 to 0.7 μm, 0.9 to 1.2 μm, and 1.9 to 2.5 μm; monolithic, stable laser arrays with output powers greater than 1 W; and low threshold (<1 mA) lasers with a modulation capability of 10 GHz.

REFERENCES

1. K. Hamada, M. Wada, H. Shimizu, M. Kume, A. Yoshikawa, F. Tajira, K. Itoh, and G. Kano, "A 0.2 W CW Laser with Buried Twin-Ridge Substrate Structures," *Ninth IEEE International Semiconductor Laser Conference*, August 7–10, 1984, Rio de Janeiro, Brazil.

2. B. Goldstein, M. Ettenberg, N. A. Dinkel, and J. K. Butler, "A High-Power Channeled-Substrate-Planar AlGaAs Laser," *Applied Physics Letters*, Vol. 47, No. 7, 1 October 1985, pp 655–57.
3. D. R. Scifres, C. Lindstrom, R. D. Burnham, W. Streifer, and T. L. Paoli, "Phase-Locked GaAlAs Laser Diode Emitting 2.6 W CW from a Single Mirror," *Electron. Letts.*, *19*, 169 (1983).
4. D. Botez, J. C. Connolly, M. Ettenberg, and D. B. Gilbert, "Very High CW Output Power and Power Conversion Efficiency from Current-Confined CDH-LOC Diode Lasers," *Electron. Letts.*, *19*, 21, pp. 882–83, Oct. 1983.
5. I. Mito, M. Kitamura, K. E. Kobayashi, S. Murato, M. Seki, Y. Odagiri, H. Nishimoto, M. Yamaguchi, and K. O. Kobayashi, "InGaAsP Double-Channel-Planar-Buried-Heterostructure Laser Diode (DC-PBH LD) with Effective Current Confinement," *IEEE J. Lightwave Technol.*, *LT-1*, pp. 195–202 (1983).
6. R. L. Hartman, N. E. Schumaker, and R. W. Dixon, "Continuously Operated (AlGa)As Double-Heterostructure Lasers with 70°C Lifetimes As Long As Two Years," *Appl. Phys. Lett.*, *31*, pp. 756–59 (1977).
7. R. L. Hartman and R. W. Dixon, "Reliability of DH GaAs Lasers at Elevated Temperatures," *Appl. Phys. Lett.*, *26*, pp. 239–42 (1975).
8. M. Ettenberg and H. Kressel, "The Reliability of (AlGa)As CW Laser Diodes," *IEEE J. Quantum Electron.*, *QE-16*, pp. 186–96 (1980).
9. J. E. Bowers, T. L. Koch, B. R. Hemerway, T. J. Bridges, E. G. Burkhardt, and D. P. Wilt, "8-GHz Bandwidth 1.52 μm Vapor Phase Transported InGaAsP Lasers," *Digest of Technical Papers*, Conference on Lasers and Electro-Optics, Baltimore, MD, May 21–24, 1985.
10. J. E. Bowers, B. R. Hemerway, A. H. Gnark, T. J. Bridges, E. G. Burkhardt, D. P. Wilt, and S. Maynard, "High Frequency Constricted Mesa Lasers," *Appl. Phys. Lett.*, *47*, July 15, 1985.
11. G. H. B. Thompson, *Physics of Semiconductor Laser Devices* (New York: John Wiley and Sons, 1980).
12. H. Kressel and J. K. Butler, *Semiconductor Lasers and Heterojunction LED's*, New York: Academic Press, 1977).
13. R. Lang, "Horizontal Mode Deformation and Anomalous Lasing Properties of Stripe Geometry Injection Lasers—Theoretical Model," *Jpn. J. Appl. Phys.*, *16*, pp. 205–206 (1977).
14. N. Chinone, "Nonlinearity in Power Output-Current Characteristics of Stripe Geometry Injection Lasers," *J. Appl. Phys.*, *48*, pp. 3237–43 (1977).
15. D. B. Carlin, private communication.
16. T. Tsukada, "GaAs-GaAl$_x$As Buried-Heterostructure Injection Lasers," *J. Appl. Phys.*, *45*, pp. 4899–906 (1974).
17. K. Aiki, "Transverse Mode Stabilized Al$_x$Ga$_{1-x}$As Injection Lasers with Channeled-Substrate-Planar Structure," *IEEE J. Quantum Electron.*, *QE-14*, pp. 89–94 (1978).
18. H. Namizaki et al. "Transverse-Junction-Stripe Stripe-Geometry Double-Heterostructure Lasers with Very Low Threshold Current," *Appl. Phys.*, *45*, pp. 2785–86 (1974).

19. D. Botez, "Single-Mode CW Operation of 'Double-Dovetail' Constricted DH (AlGa)As Diode Lasers," *Appl. Phys. Lett.*, *33*, pp. 872–74 (1978).
20. J. Manning and R. Olshansky, "The Carrier-Induced Index Change in AlGaAs and 1.3 μm InGaAsP Diode Lasers," *IEEE J. Quantum Electron.*, *QE-19*, 10, pp. 1525–30, October 1983.
21. G. Clarke, L. Heflinger, and C. Roychoudhuri, TRW IR&D Report #8100 4342.
22. W. T. Tsang, N. A. Olson, and R. A. Logan, "High-Speed Direct Single-Frequency Modulation with Large Tuning Rate and Frequency Excursion in Cleaved-Coupled-Cavity Semiconductor Lasers," *Appl. Phys. Lett.*, *42*, 8, pp. 650–52, April 15, 1983.
23. K. H. Cameron, P. J. Chidgey, and K. R. Preston, "102 km Optical Fibre Transmission Experiments at 1.52 μm Using an External Cavity Controlled Laser Transmitter Module," *Electron. Letts.*, *18*, 15, pp. 650–51, July 22, 1982.
24. Y. Suematsu, S. Arai, and K. Kishino, "Dynamic Single-Mode Semiconductor Lasers with a Distributed Reflector," *J. Lightwave Tech.*, *LT-1*, 1, pp. 161–76, March 1983.
25. Y. Tohmori, Y. Suematsu, H. Tsushima, and S. Arai, "Wavelength Tuning of GaInAsP/InP Integrated Laser with Built-In Distributed Bragg Reflector," *Electron. Letts.*, *19*, 17, pp. 656–57, August 18, 1983.
26. Y. Sakakibra, K. Furuya, K. Utaka, and Y. Suematsu, "Single-Mode Oscillation under High-Speed Direct Modulation in GaInAsP/InP Integrated Twin-Guide Lasers with Distributed Bragg Reflectors," *Electron. Letts.*, *16*, 12, pp. 456–58, June 5, 1980.
27. M. Yamaguchi, M. Kitamura, I. Mito, S. Murata, and K. Kobayashi, "Highly Efficient Single Longitudinal-Mode Operation of Antireflection-Coated 1.3 μm DFB-DC PBH LD," *Electron. Lett.*, *20*, 6, pp. 233–35, March 15, 1984.
28. J. M. Hammer, C. C. Neil, N. W. Carlson, M. T. Duffy, and J. M. Shaw, "Single-Wavelength Operation of the Hybrid-External- Bragg-Reflector-Waveguide Laser under Dynamic Conditions," *Appl. Phys. Lett.*, *47*, 7 pp. 183–85, August 1, 1985.
29. E. Kapan, C. Lindsey, J. Katz, S. Margolit and A. Yariv, "Coupling Mechanism of Gain-Guided Integrated Semiconductor Laser Arrays," *Appl. Phys. Lett.*, *44*, 4, pp. 389–91, February 15, 1984.
30. D. E. Ackley, "Phase-Locked Injection Laser Arrays with Non-Uniform Stripe Spacing," *Electron. Lett.*, *20*, 17, pp. 695–97, August 16, 1984.
31. D. Botez and J. C. Connolly, "High Power Phase Locked Arrays of Index Guided Diode Lasers," *Applied Physics Letters*, *43*, pp. 1096–98, December, 1983.
32. W. T. Tsang, "Extremely Low Threshold (AlGa)As Graded-Index Waveguide Separate-Confinement Heterostructure Lasers Grown by Molecular Beam Epitaxy," *Appl. Phys. Lett.*, *40*, 3, pp. 217–19, February 1, 1982.
33. S. Yamamato, H. Hayashi, T. Hayakawa, N. Miyauchi, S. Yano and T. Hijikata, "Room-Temperature cw Operation in the Visible Spectral Range of 680–700 nm by AlGaAs Double Heterojunction Lasers," *Appl. Phys. Lett.*, *41*, 9, pp. 796–98, November 1, 1982.
34. K. Uqmi, S. Nakatsuka, T. Ohtoshi, Y. Ono, N. Chinone, and T. Kajimura, "Stable Fundamental Transverse-Mode Operation of Index-Guided Visible GaAs/

GaAlAs Multiquantum-Well Lasers," *9th IEEE International Semiconductor Laser Conference*, Rio de Janeiro, Brazil, August 1984.

35. I. Ladany and H. Kressel, "Degradation in Short Wavelength (AlGa)As Light-emitting Diodes," *Electronics Letters*, *14*, pp. 407–9, June 1978.
36. H. Shimizu, K. Itoh, M. Wada, T. Sugino, and I. Teramoto, "Improvement in Operation Lives of GaAlAs Visible Lasers by Introducing GaAlAs Buffer Layers," *IEEE J. Quantum Electron.*, *QE-17*, 5, pp. 763–67, May 1981.
37. M. Ettenberg, "A Statistical Study of the Reliability of Oxide-Defined Stripe CW Lasers of (AlGa)As," *J. Appl. Phys.*, *50*, pp. 1195–1202, (1979).
38. W. B. Joyce, R. W. Dixon, and R. L. Hartman, "Statistical Characterization of the Lifetimes of Continuously Operated (AlGa)As Doule Heterostructure Lasers," *Appl. Phys. Lett.*, *28*, pp. 648–86 (1976).
39. *AT&T Technical Journal*, Special Issue on Assuring High Reliability of Lasers and Photodetectors for Submarine Lightwave Cable Systems, *64*, 3, May 1985.
40. D. S. Peck and C. H. Zierdt, "The Reliability of Semiconductor Devices in the Bell System," *Proc. IEEE*, *62*, pp. 185–211 (1974).
41. C. Lanza, K. L. Konnerth, and C. E. Kelly, "Aging Effects in GaAs Electroluminescent Diodes," *Solid-State Electron.*, *10*, pp. 21–31 (1967).
42. H. Kressel, M. Ettenberg, and H. F. Lockwood, "Effects of Edges on the Reliability of GaAs and (AlGa)As Heterojunction LEDs," *J. Electron. Mat.*, *6*, pp. 467–81 (1977).
43. N. E. Byer, "Role of Optical Flux and of Current Density in Gradual Degradation of GaAs Injection Lasers," *IEEE J. Quantum Electron.*, *QE-5*, pp. 242–45 (1969).
44. D. H. Newman and S. Ritchie, "Gradual Degradation of GaAs Double-Heterostructure Lasers," *IEEE J. Quantum Electron.*, *QE-9*, p. 300 (1973).
45. K. Mizuishi et al., "Acceleration of the Gradual Degradation in (GaAl)As Double-Heterostructure Lasers as an Exponent of the Value of the Driving Current," *J. Appl. Phys.*, *50*, pp. 6669–74 (1979).
46. R. T. Lynch, "Effect of Screening Tests on the Lifetime Statistics of Injection Lasers," *IEEE J. of Quantum Electron.*, *QE-16*, 11, pp. 1244–47, Nov. 1980.
47. G. Gordon, R. Nash, and R. Hartman, "Purging: A Reliability Assurance Technique for New Technology Semiconductor Devices," *IEEE Trans. on Electronic Devices*, *EDL4*, 12, Dec. 1983.
48. R. H. Saul and F. S. Chen, "Reliability Assurance for Devices with a Sudden Failure Characteristic," *IEEE Trans. on Electronic Devices*, *EDL4*, 12, Dec. 1983.
49. D. A. Shaw and P. R. Thornton, "Catastrophic Degradation in GaAs Laser Diodes," *Solid State Electron.* *13*, pp. 919–24 (1970).
50. N. Chinone, R. Ito, and O. Nakada, "Limitations of Power Outputs from Continuously Operating GaAs-$GA_{i-x}Al_xAs$ Double-Heterostructure Lasers," *J. Appl. Phys.*, *47*, pp. 785–86 (1976).
51. B. W. Hakki, and F. R. Nash "Catastrophic Failure in GaAs Double Heterostructure Injection Lasers," *J. Appl. Phys.*, 45, pp. 3907–15 (1974).
52. T. Yuasa et al., "Degradation of (AlGa)As Lasers Due to Facet Oxidation," *Appl. Phys. Lett.*, *32*, pp. 119–120 (1978).

53. T. Suzuki and M. Ogawa, "Degradation of Photoluminescence Intensity Caused by Excitation Enhanced Oxidation of GaAs Surfaces," *Appl. Phys. Lett.*, *31*, pp. 473–74 (1977).

54. Y. Kadota, K. Chino, Y. Onodera, H. Namizaki, and S. Takamiya, "Aging Behavior and Surge Endurance of 870-900 nm AlGaAs Lasers with Nonabsorbing Mirrors," *IEEE J. Quantum Electron.*, *QE-20*, 11, pp. 1247–51, Nov. 1984.

55. H. Blauvelt, S. Margolit, and A. Yariv, "Large Optical Cavity AlGaAs Buried Heterostructure Window Lasers," *Appl. Phys. Lett.*, *40*, 12 pp. 1029–31, June 15, 1982.

56. D. Botez and J. C. Connolly, "Non-Absorbing-Mirror (NAM) CDH-LOC Diode Lasers," *Electronics Letters*, *20*, 13, pp. 530–32, June 21, 1984 and Erratum, *Electronics Letters*, *20*, 17, p. 710, August 16, 1984.

57. S. Yamamoto, H. Hayashi, T. Hayakawa, N. Miyauchi, S. Yano, and T. Hijikata, "High Optical Power cw Operation in Visible Spectral Range by Window V-Channeled Substrate Inner Stripe Lasers," *Appl. Phys. Lett.*, *42*, 5, pp. 406–8, March 1983.

58. J. A. F. Peek, "Water Vapor, Facet Erosion and the Degradation of (Al,Ga)As DH Lasers Operated at CW Output Powers up to 3 mW/μm Stripe Width," *IEEE J. Quantum Electron.*, *QE-7*, p. 781 (1981).

59. B. C. DeLoach, Jr., et al., "Degradation of DW GaAS Double-Heterojunction Lasers at 300 K," *Proc. IEEE*, *61*, pp. 1042–44 (1973).

60. P. Petroff and R. L. Hartman, "Rapid Degradation Phenomenon in Heterojunction GaAlAs-GaAs Lasers," *J. Appl. Phys.*, *45*, pp. 3899–903 (1974).

61. P. W. Hutchinson and P. S. Dobson, "Defect Structure of Degraded GaAlAs-GaAs Double Heterojunction Lasers," *Philos. Magazine*, *32*, pp. 745–54 (1975).

62. P. W. Hutchinson et al., "Defect Structure of Degraded Heterojunction GaAlAs-GaAs Lasers," *Appl. Phys. Lett.*, *26*, pp. 250–52 (1975).

63. C. C. Shen, J. J. Hsieh, and T. A. Lind, "1500-h Continuous CW Operation of Double-Heterostructure GaInAsP/InP Lasers," *Appl. Phys. Lett.*, *30*, pp. 353–54 (1977).

64. M. Ettenberg and C. J. Nuese, "Reduced Degradation in $InGa_{1-x}A_xAs$ Electroluminescent Diodes," *J. Appl. Phys.*, *46*, 5, pp. 2139–42, May 1975.

65. Y. Seki, I. Matsui, and H. Watanabe, "Impurity Effect on the Growth of Dislocation-Free InP Single Crystals," *J. Appl. Phys.*, *47*, 7, pp. 3374–76, July 1976.

66. G. R. Woolhouse, A. E. Blakeslee, and K. K. Shih, "Detection and origins of Crystal Defects in GaAs/GaAlAs LPE Layers," *J. Appl. Phys.*, *47*, pp. 4349–52 (1976).

67. M. Ishii et al., "Suppression of Defect Formation in GaAs Layers by Removing Oxygen in LPE," *Appl. Phys. Lett.*, *29*, pp. 375–76 (1976).

68. A. R. Goodwin et al., *IEEE Semiconductor Laser Conf.*, Nemu-No-Sato, Japan, September 1976.

69. H. Kressel, M. Ettenberg, and I. Ladany, "Accelerated Step Temperature Aging of $Al_xGa_{1-x}As$ Heterojunction Laser Diodes," *Appl. Phys. Lett.*, *32*, pp. 305–8 (1978).

70. S. Ritchie et al., "The Temperature Dependence of Degradation Mechanisms in Long Lived (AlGa)As DH Lasers," *J. Appl. Phys.*, *49*, pp 3127–32 (1978).

71. K. Fugiwara et al., "Aging Characteristics of $Ga_{1-x}Al_xAs$ Double Heterostructure Lasers Bonded with Gold Eutectic Alloy Solder," *Appl. Phys. Lett.*, *34*, pp. 668–70 (1979).

72. T. Kobayash and G. Iwane, "Three Dimensional Thermal Problems of Double Heterostructure Semiconductor Lasers," *J. Appl. Phys.*, *16*, pp. 1403–8, (1977).

73. H. D. Wolf, K. Mettler, and K. H. Zschauer, *Japan J. Appl. Phys.*, *20*, 9, L693 (1981).

74. R. Salathe et al., "Laser-Alloyed Stripe-Geometry DH Lasers," *Appl. Phys. Lett.*, *35*, pp. 439–41 (1979).

75. R. L. Mozzi, W. Fabian, and F. J. Piekarski, "Nonalloyed Ohmic Contacts to N-GaAs by Pulse-Electron-Beam Annealed Selenium Implants," *Appl. Phys. Lett.*, *35*, pp. 336–39 (1979).

76. T. L. Paoli, "Changes in the Optical Properties of CW (AlGa)As Junction Lasers During Accelerated Aging," *IEEE J. Quantum Electron.*, *QE-13*, pp. 351–59 (1977).

77. D. J. Channin, M. Ettenberg, and H. Kressel, "Self-Sustained Oscillations in (AlGa)As Oxide-Defined Stripe Lasers." *Journal of Applied Physics, 50*, No. 11, Part 1, pp. 6700–706, November 1979.

78. K. Mizuishi et al., "The Effect of Sputtered SiO_2 Facet Coating Films on the Suppression of Self Sustained Pulsations in the Output of (GaAl)As Double Heterostructure Lasers During CW Operation," *IEEE J. Quantum Electron.*, *QE-16*, pp. 728–34 (1980).

79. N. Chinone, K. Aiki, and R. Itop, "Stabilization of Semiconductor Laser Outputs by a Mirror Close to a Laser Facet," *Appl. Phys. Lett.*, *33*, pp. 990–92 (1978).

80. M. Dixon and B. A. Dean, "Aging of the Light-Current Characteristic of Proton-Bombarded AlGaAs Lasers Operated at 30C in Pulsed Conditions," *3rd International Conference on Integrated Optics and Optical Fiber Communication*, San Francisco, CA, April 27–29, 1981.

81. D. J. Channin, D. Botez, C. C. Neil, J. C. Connoly, and D. W. Bechtle, "Modulation Characteristics of Constricted Double-Heterojunction AlGaAs Laser Diodes," *J. Lightwave Tech.*, *LT-1*, 1, pp. 146–61, March 1983.

82. T. Ikegami and Y. Suematsu, "Direct Modulation of Semiconductor Junction Lasers," *Electron. Commun. Japan*, *B51*, pp. 51–58 (1968).

83. T. L. Paoli and J. E. Ripper, "Direct Modulation of Semiconductor Lasers," *Proc. IEEE*, *58*, pp. 1457–65 (1970).

84. K. Y. Lau and A. Yariv, "Ultra-High Speed Semiconductor Lasers," *IEEE J. of Quantum Electron.*, *QE-21*, 2, pp. 121–38, February 1985.

85. D. J. Channin, D. Redfield, and D. Botez, "Effect of Injection-Current Confinement on Modulation of CDH-LOC AlGaAs Laser Diodes," *Conference Proceedings of the Ninth IEEE International Semiconductor Laser Conference*, Rio de Janeiro, Brazil, August 7-10, 1984.

86. D. M. J. Compton and R. A. Cesena, "Mechanisms of Radiation Effects on Lasers," *IEEE Trans. Nucl. Sci.*, *NS-14*, 6, p. 16 (1967).

87. C. E. Barnes, "Radiaton Effects in Electroluminescent Diodes," *IEEE Trans. Nucl. Sci.*, *NS-18*, 6, p. 322 (1971).
88. C. E. Barnes, "Neutron Damage in Epitaxial GaAs Laser Diodes," *J. Appl. Phys.*, *42*, p. 194 (1971).
89. C. E. Barnes, "Neutron Damage in GaAs Laser Diodes: At and Above Laser Threshold," *IEEE Trans. Nucl. Sci.*, *NS-19*, pp. 382–85 (1972).
90. C. E. Barnes, "Increased Radiation Hardness of GaAs Laser Diodes at High Current Densities," *J. Appl. Phys.*, *45*, p. 3485 (1974).
91. H. J. Minden, "Effects of Proton Bombardment on the Properties of GaAs Laser Diodes," *J. Appl. Phys.*, 47, p. 1090 (1976).
92. J. O. Schroeder, B. W. Noel, and H. D. Southward, "Radiation Damage Induced Time Delay in GaAs Lasers," *IEEE Trans. Nucl. Sci.*, *NS-20*, 6, p. 261 (1973).

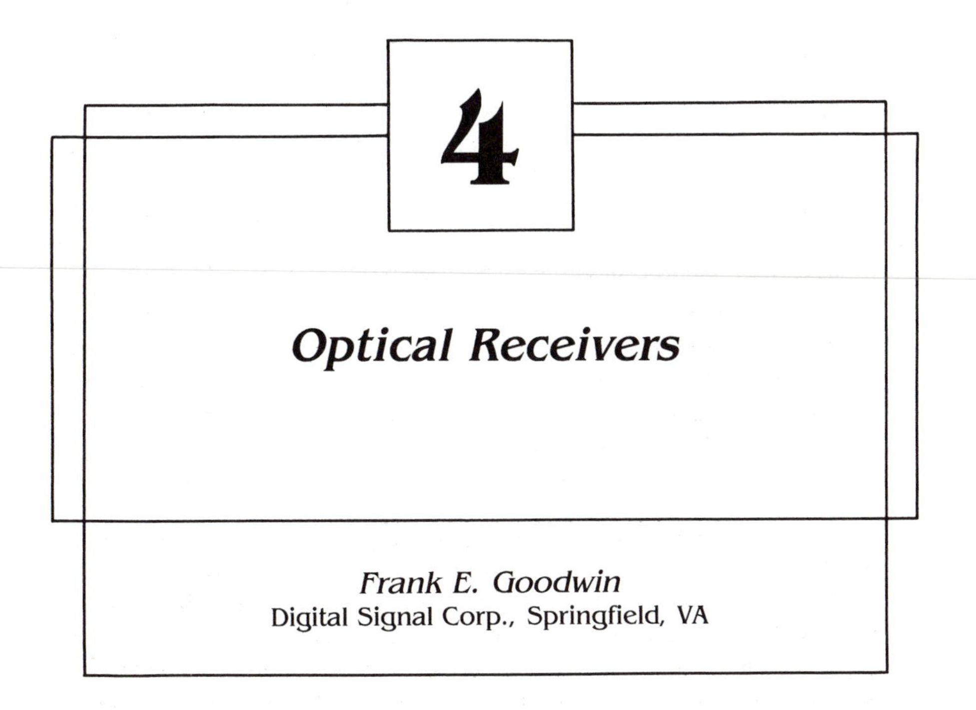

4

Optical Receivers

Frank E. Goodwin
Digital Signal Corp., Springfield, VA

4.1 INTRODUCTION

A laser signal is one that contains information. Detection of the mere presence of a signal is a bit of information. A laser beam may be modulated and contain information to a very high data rate. The quality of the modulation and coding are crucial to the efficiency of the link, and several optimal modulation and coding schemes have been developed. However, it is the detection process which mandates how the laser should be modulated. For example, direct (or quantum) detection suggests that the laser be pulsed (PAM) while coherent or heterodyne detection suggests that frequency or phase modulation be used.

It is instructive to begin a discussion of optical signal detection with the physics of detection and a review of the different ways there are to detect the presence of light:

- **Photo absorption**—Heating by absorption of the light produces a detectable potential across a thermocouple or a piezoelectric element; a naturally slow process.
- **Photo emission**—Photons exchange kinetic energy with electrons that are emitted from a photocathode surface, multiplied by secondary emission, and detected electronically, a process that is moderately fast.
- **Photo conduction**—Photons give up energy in a semiconductor by

creating electron-hole pairs thereby increasing the conductivity of the material, a process that is generally slow but which can be made moderately fast.

- **Photo voltaic**—Photons give up energy by elevating electrons to a higher energy level thereby causing the generation of a potential across a diode junction, moderately fast but limited by capacitive effects.
- **Biased photodiode**—Capacitance is reduced by back-biasing a photodiode and operating it in a quasi photo conductive mode; can be made to be extremely fast.
- **Avalanche photodiode**—Back-bias potential is set at the threshold of breakdown where photoelectron events experience current multiplication due to the avalanche effect, a moderately fast process.

Laser communications systems of the future will most certainly use one of the three latter photodetection processes because of their speed and simplicity. The remainder of this chapter will be devoted to the discussion of their use as direct detection systems and as coherent detection systems. Direct detection is commonly referred to as quantum counting since an electron may be produced for each photon event. The phase information contained in the optical signal is lost in the direct detection process. On the other hand, coherent detection preserves the phase information by imposing a local oscillator laser beam on the signal prior to detection. A heterodyne or homodyne process is thereby produced.

Direct detection of optical signals is practical with photoemissive devices such as photomultipliers and avalanche photodiodes where large intrinsic gain permits the generation of large numbers of photoelectrons for each photon incident on the photosurface. Photoemissive devices, however, roll off at about 1 μ because the energy of the individual photons at the longer wavelengths cannot overcome the work function of the photocathode surface. From about 1.1 μm out to the far infrared, the best optical detectors are semiconductor photodiodes. Direct detection by these photodiodes is usually limited by thermal (amplifier) noise. At the longer wavelengths, especially at 10 μ, the use of coherent detection has a beneficial effect; the mixing process of the signal with the local oscillator produces a conversion gain sufficient to override the effects of thermal (amplifier) noise.

4.2 DIRECT DETECTION

Direct detection is defined as the direct use of electrons produced by ionizations of photons incident on the detector. For example, assume an optical flux ϕ shown in Figure 4-1, $\phi = P/h\nu$ where P is the optical power and $h\nu$ is the energy per photon. T_{eff} is the effective temperature of the load resist-

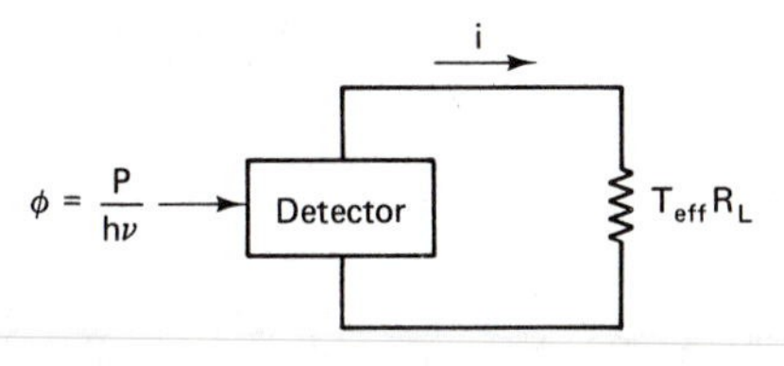

Figure 4-1 Direct Detection Equivalent Circuit

ance. R_L is the load resistance. A signal current is produced in the detector of

$$i_s = \eta Ge \left(\frac{P_s}{h\nu}\right)$$

where

η = number of ionizations per photon (quantum efficiency)
G = number of electrons per ionization (gain)
e = electronic charge

The gain of a photomultiplier may be 10^5 or higher, where a single photon can generate thousands of electrons and where individual photon events are easily detected. An avalanche photodiode may have a gain of 10^2 or more. However, most infrared photodiodes do not have intrinsic gain.

The optical power P above may be composed of a signal power P_s and a background power P_b. In such a case, the total current is

$$i = \frac{\eta Ge(P_s + P_b)}{h\nu} + i_d \tag{4-1}$$

The power in the resistor is given by (neglecting dark current)

$$i^2R_L = \frac{\eta^2G^2e^2(P_s + P_b)^2R_L}{(h\nu)^2} \tag{4-2}$$

The noise power (shot) in the resistor is the shot noise produced by the direct current in the detector.

$$i_n^2R_L = 2(Ge)i\,BR_LF \tag{4-3}$$

F is the noise factor defined by the increase in noise introduced by the current gain process and B is the bandwidth. The value of F varies for different types of detectors.

F = 1 for photodiodes
F = 1.3 for photomultipliers
F = 2 for photoconductors
F > 2 for avalanche photodiodes

The load resistor also has thermal noise so that the total noise power in the load resistor is

$$P_n = 2(Ge)\, i_{dc}\, BRF + 4\, kTB. \tag{4-4}$$

Now i_{dc} in the above equation is the total dc current in the circuit and is the sum of that produced by the signal flux, the background flux, and dark current. In this discussion, we are assuming the dark current to be negligible and that the main shot current is produced by the signal and the background flux

$$i_{dc} = \eta Ge \frac{P_s + P_b}{h\nu}$$

The noise power in the load resistor becomes

$$P_n = \frac{2\eta(Ge)^2\,(P_s + P_b)\, BRF}{h\nu} + 4\, kTB \tag{4-5}$$

4.2.1 Signal to Noise

Thermal Limited Noise Equivalent Power $(NEP)_T$

NEP is defined as the amount of signal power spectral density (per root hertz of bandwidth) necessary to make the signal to noise ratio in an optical receiver equal to unity. The NEP can be dominated by amplifier noise (thermal), background noise, dark current, or by the noise on the signal itself.

The NEP of a solid-state detector is usually dominated by the thermal noise of the amplifier circuit. Although the detector quantum efficiency is near unity, the electronic noise of the amplifier is many times greater than the signal current generated by a photoelectron. The signal current can be written:

$$i_s = \eta Ge \frac{(NEP)_T\, B}{h\nu}$$

and the noise current is

$$i_n = \frac{4\, kTB}{R}$$

Setting $(i_s/i_n)^2 = 1$, the thermal limited NEP is given by

$$(NEP)_t \sqrt{B} = \frac{h\nu}{\eta Ge}\sqrt{\frac{4\, kTB}{R}} \tag{4-6}$$

For the equivalent thermal current to be small compared with quantum noise, the gain of the photodetector must be very large. For a typical case of a photomultiplier followed by an amplifier having a 3-dB noise figure and an input resistance of $10^4\ \Omega$,

$$\frac{h\nu}{\eta Ge}\sqrt{\frac{4\,kT}{R}} << \frac{h\nu}{\eta} \qquad G >> \sqrt{\frac{4\,kT}{e^2R}}$$

where $G >> 10^7$.

Background Limited Noise-Equivalent Power $(NEP)_B$

For the background limited case, the signal current must be greater than the shot noise produced by the dc background flux. For a background flux of $P_b/h\nu$, the dc current produced in the detector is

$$i_{dc} = \eta Ge\frac{P_b}{h\nu}$$

and noise current is the shot noise introduced by the same i_{dc}

$$i_n = \sqrt{2e\ i_{dc}\ B}$$

Again, the signal current defines the background limited $(NEP)_B$ by

$$i_s = \eta Ge\left[\frac{(NEP)_B\sqrt{B}}{h\nu}\right]$$

Setting $(i_s/i_n)^2 = 1$ and solving for $(\text{NEP})_B$, we have

$$(NEP)_B\sqrt{B} = \frac{h\nu}{\eta Ge}\sqrt{2e\ i_{dc}B}$$

$$(NEP)_B\sqrt{B} = \sqrt{2(h\nu B/\eta)\ P_B}, \qquad \frac{h\nu B}{\eta} = P_q$$

$$(NEP)_B\sqrt{B} = \sqrt{2\ P_qP_b} \tag{4-7}$$

Signal Shot-Noise Limited Noise Equivalent Power $(NEP)_S$

The case where signal strength is great or where background and thermal noise is negligible, the noise in the system is determined by signal shot noise. The computation of the signal shot noise is similar to that for background noise where average signal flux is substituted for background flux. The dc current produced in the detector by the average signal current is

$$i_{dc} = \eta Ge\frac{P_s + P_b}{h\nu}$$

where now as we have said $P_s >> P_b$.

The signal shot noise power $(NEP)_S$ is thus

$$(NEP)_S = \sqrt{2P_qP_s} \tag{4-8}$$

Signal-to-Noise Ratio of Unmodulated Carriers $(S/N)_{dc}$

In this section, we have described the types of noise encountered in direct detection of optical signals. We now define what is meant by the signal-to-noise ratio. We define the $(S/N)_{\mathrm{dc}}$ as that for the detection of unmodulated carriers,

$$(S/N)_{\mathrm{dc}} = \left(\frac{P_s}{NEP\sqrt{B}}\right)^2 \tag{4-9}$$

where P_s is the actual optical signal power and NEP is the hypothetical equivalent optical noise power for the three cases described in Equations (4-6) through (4-8). Accordingly, these are:

Thermal limited

$$\left(\frac{S}{N}\right)_{\mathrm{dc}} = \left[\frac{\eta Ge}{h\nu}\sqrt{\frac{R}{4\,kTB}}\,P_s\right]^2$$

Background limited

$$\left(\frac{S}{N}\right)_{\mathrm{dc}} = \left[\frac{P_s}{2\sqrt{P_q P_b}}\right]^2$$

Signal shot limited

$$\left(\frac{S}{N}\right)_{\mathrm{dc}} = \left[\frac{P_s}{2\sqrt{P_q P_s}}\right]^2 = \eta\,\frac{P_s}{2h\nu B} \tag{4-10}$$

General

$$\left(\frac{S}{N}\right)_{\mathrm{dc}} = \frac{P_s^2}{\left(\frac{h\nu}{\eta Ge}\sqrt{\frac{4\,kTB}{R}}\right)^2 + \left(\sqrt{2\,P_q\,(P_b + P_s)}\right)^2}$$

The detection of single pulse modulation does not directly depend on the NEP of the detector, but rather on the number of photoelectrons necessary for a given detection probability. The $(S/N)_m$ associated with this requirement may be expressed as

$$(S/N)_m = \left[\eta\,\frac{E_s}{h\nu} \div \frac{\eta}{h\nu}\sqrt{2E_s E_q}\right] = \frac{E_s}{2E_q} = \frac{K_s}{2} \tag{4-11}$$

where E_s is the signal energy.

K_s is the number of photoelectrons produced per pulse. Typically, for quantum limited detection and high quantum efficiencies, $K_s = 15$ for a probability of error of 10^{-5}.

4.2.2 Demodulation and Information Signal-To-Noise Ratio

Message or information signal-to-noise ratio can be estimated from

$$\left(\frac{S}{N}\right)_m = \left(\frac{P_m}{NEP}\right)^2 \tag{4-12}$$

where P_m is the amount of power in the information sidebands of the signal. If the modulation/coding scheme is optimal (one in which all of the transmitted energy contains information), P_m is equal to P_s and the information signal to noise ratio is exactly the same as that of the unmodulated carrier. Not all modulation schemes are optimal, however. The efficiency of the modulation/coding scheme is estimated by the ratio of message power P_m to total carrier power. Efficiencies of several suboptimal and optimal schemes are listed in the following table:

Modulation	P_m/P_s
100% Intensity modulation (analog)	0.33
100% Subcarrier IM	0.11
100% Polarization modulation (analog)	0.64
Pulse amplitude modulation (PAM)	1.00

It is seen that the information or message signal-to-noise ratios for typical intensity modulation systems is significantly less than the dc signal-to-noise ratios. The exception is pulse modulation where all the signal contains information. Here, the message signal-to-noise ratio from Equation (4-11) is proportional to the number of photoelectrons produced per pulse, K_s.

$$\left(\frac{S}{N}\right)_m = \eta\frac{K_s}{2} = \eta\frac{P_s}{2h\nu b}$$

and, to put the expression in the form of video bandwidth, the relation between bit rate, b, and minimum bandwidth, B, is $B = 0.7b$.
Thus, the message signal to noise for pulse modulation is:

$$\left(\frac{S}{N}\right)_m = \eta\frac{P_s}{2.8\,h\nu B} \tag{4-13}$$

4.3 ILLUSTRATIVE EXAMPLE OF A PULSE LASER RECEIVER

The examples given thus far were ideal or special cases that were chosen to illustrate a point. Actual cases are almost always a combination of conditions that can be fairly difficult to analyze. Through necessity, there has evolved a common ground for all direct detection systems; the two key parameters that best represent the system are bit error rate (BER) and photoelectrons per bit (K_s). The illustrative example that is presented here is for a receiver having the following characteristics:

Data rate (b)	1.0 Mbps
Wavelength	8500 Å (AlGaAs)
Detector	Silicon APD with noise factor of $F = 2 + 0.01G$ where G = APD gain and quantum efficiency = 0.8.
Preamplifier	High impedance FET, $R = 50{,}000\ \Omega$ $C = 5$ pF
Background	Three cases: solar $K_b = (6.8 \times 10^{10})$ n/b lunar $K_b = (6.8 \times 10^{7})$ n/b dark $K_b = 0$.

Defining signal current as $i_s = GebK_s$, Equation (4-10) can now be written as

$$\left(\frac{S}{N}\right)_m = \frac{K_s^2}{\dfrac{K_{t'}}{G^2} + 2F(K_s + K_b)} \tag{4-14}$$

where

$$K_{t'} = \frac{4\,kT}{e^2 R_L} \times \frac{1}{b}$$

attributable to the preamplifier noise. The total noise equivalent "counts" per bit is

$$K_n = \sqrt{\frac{K_{t'}}{G^2} + 2F(K_s + K_b)}$$

The value of APD gain can be optimized to produce a maximum signal-to-noise ratio. Let the APD noise factor F = 2 + 0.01G. Taking the first derivative of (S/N) with respect to B, setting it to zero, and solving for G gives the optimum gain for the APD,

$$(G)_{\text{opt}} = \sqrt[3]{\frac{2K_t'}{0.01\,(K_s + K_b)}} \tag{4-15}$$

Notice that the optimum gain G may differ widely according to the values of thermal noise and background noise.

Bit error rate can be estimated as a function of (S/N) or signal and noise counts from a number of models. A thorough discussion of bit-error rate deserves a separate volume. The model used here is not rigorous and is intended only as an example. Assume that the bit error rate BER is

$$BER = \frac{1 - \text{ERFX}}{2}$$

where (4-16)

$$X = \frac{1}{2\sqrt{2}}\sqrt{\frac{S}{N}}$$

The BER equation is the last element required to produce a BER versus signal power relationship. However, the computation is complex and iterative. The process involves the following steps:

1. Estimate (S/N) for required BER [Equation (4-16)].
2. Compute background counts K_b for assumed background.
3. Compute thermal counts K_t' for assumed receiver parameters.
4. Estimate noise factor F for an assumed APD gain of 100.
5. Compute signal counts K_s for parameters 1 through 4.
6. Compute optimum APD gain [Equation (4-15)] G for parameters 1 through 5.
7. Repeat steps 3 through 6 using new value of G until value of optimum gain G stabilizes.

The process is repeated for each new value of BER and a graphic relation can be plotted. Figure 4-2 illustrates an example curve for a 1.00 MBps data rate where the average signal power is plotted as well as the equivalent number of signal counts per bit. A dark background was assumed. The optimum APD gain for this illustration was 147.

4.4 COHERENT DETECTION AND DEMODULATION

4.4.1 Introduction

Coherent detection is the process of mixing a local oscillator laser beam with the incoming signal beam. It has the advantage that conversion gain is achieved through the photoelectric mixing process and under optimum conditions quantum limited detection is achieved. In words, it can be described as a method of mixing two optical fields to produce an electrical current proportional to the product of the two fields. If one of these fields is the local oscillator, the effective gain can be increased arbitrarily to the point where the signal level is greater than the amplifier (thermal) noise. For more than a decade, the technique has found applications at the longer infrared wavelengths where photoemissive gain and avalanche gain is not available. In

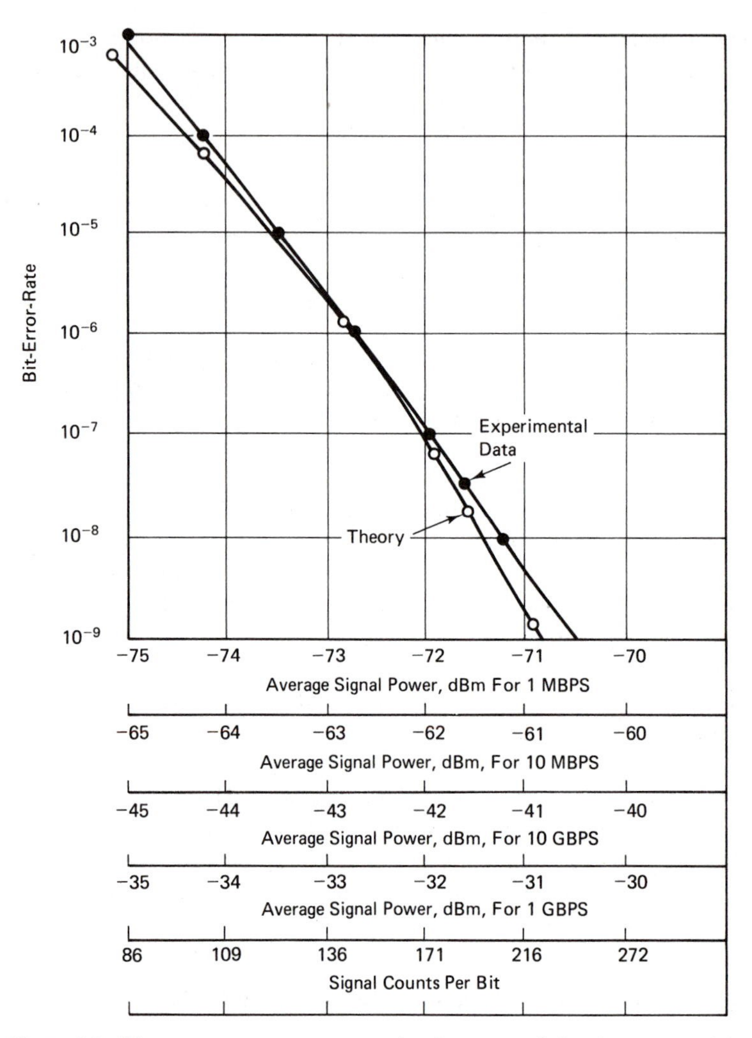

Figure 4-2 Bit-error rate versus average signal power and signal counts per bit for example communication system: Wavelength $\lambda = 8{,}500$ Å; $K_b = 0$; data rate $b = 1$, 10 Mbps and 1,104 bps; coding = PPM.

particular, the carbon dioxide laser at 10 μm offers an ideal source for use in coherent detection applications. Communications systems[1,2] and radar systems[3,4] have been developed using the technique.

Recently, coherent detection has been applied successfully to optical communications[5,6], and to fiber optical sensors[7] in the wavelength range from 0.8 to 1.5 μm. In fiber optical communication systems, a 100-fold improvement in detection sensitivity has been achieved in spite of the less than ideal quality of the laser sources at these wavelengths. In fiber optical sensors, advantage has been realized by virtue of the short optical wavelengths and

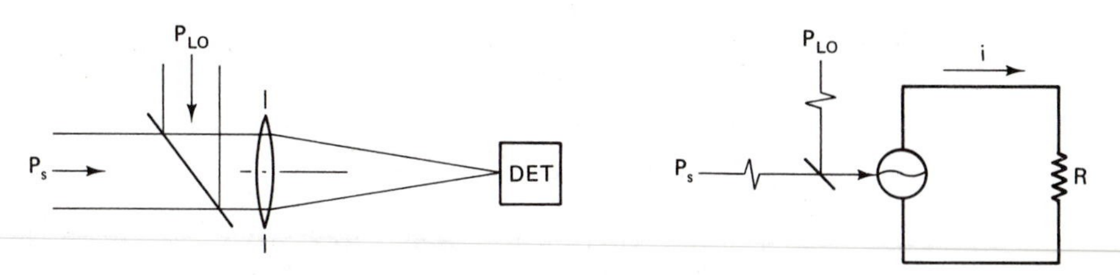

Figure 4-3 Heterodyne Detection Equivalent Circuit

the extreme sensitivity of a fiber interferometer to stress applied to the sensing element.

Use of coherent detection at any wavelength requires an understanding of the geometrical requirements of aligning local oscillator and signal beams as well as the electrical requirements of the mixing process. Section 4.4.2 presents the physical and geometrical considerations, Sec. 4.4.3 NEP versus local oscillator power, Sec. 4.4.4 heterodyne conversion gain, and Sec. 4.4.5 demodulation.

For application to fiber optics, an understanding of the frequency tunability and coherent properties of injection laser diodes is required and is presented in Sec. 4.4.6. Results achieved in optical heterodyne systems using injection laser diodes is presented in Sec. 4.4.7.

4.4.2 Physical and Geometrical Considerations

The requirements of phase matching of the wavefronts of the signal and local oscillator beam wavefronts is sufficiently important that inclusion of geometrical considerations with the physical description is necessary.[8,9,10] We therefore describe a typical signal and local oscillator configuration in Figure 4-3 with the geometric layout on the left and the electrical equivalent circuit on the right.

The distribution of the signal field E_s and the local oscillator field E_{LO} can be identical or arbitrarily different. Usually, the signal field at the detector is an Airy pattern and the local oscillator may or may not be an Airy pattern (Figure 4-4).

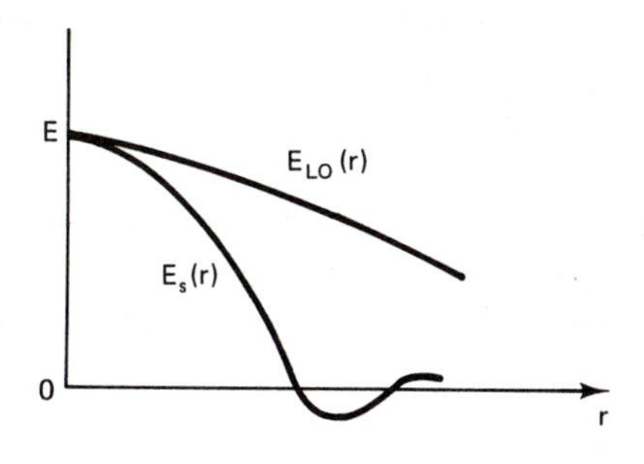

Figure 4-4 Typical field distribution of signal $E_S(V)$ and local oscillator $E_{LO}(V)$

By Poynting's theorem, the signal power and the local oscillator power can be written:

$$P_s = \frac{1}{Z_o}\int \vec{E}_s^2\, dA = \frac{1}{2Z_o}|E_s|^2 \int |U_s|^2\, dA$$

$$P_{LO} = \frac{1}{Z_o}\int \vec{E}_{LO}^2\, dA = \frac{1}{2Z_o}|E_{LO}|^2 \int |U_{LO}|^2\, dA$$

and the absolute magnitudes of the fields are

$$|E_s| = \left[\frac{2Z_o\, P_s}{|U_s|^2\, dA}\right]^{1/2},\ |E_{LO}| = \left[\frac{2Z_o\, P_s}{|U_{LO}|^2 dA}\right]^{1/2}$$

The electric field can be written as the sum of the signal and local oscillator fields

$$\vec{E} = \vec{E}_s + \vec{E}_{LO} P = \frac{1}{Z_o}\int (\vec{E}_s + \vec{E}_{LO})^2\, dA$$

and the current in the detector can be expressed as

$$i = \eta\frac{Ge}{h\nu}P = \eta\frac{Ge}{h\nu}\frac{1}{Z_o}\int (\vec{E}_s^2 + \vec{E}_{LO}^2 + 2\vec{E}_s \cdot \vec{E}_{LO})\, dA$$

The intermediate frequency i.f. current is then

$$i_{\text{i.f.}} = 2\eta\,\frac{Ge}{h\nu}\frac{|E_s|\,|E_{LO}|}{2Z_o}J\int |U_s|\,|U_{LO}|\, dA \cos \omega_{\text{i.f.}}\, t$$

$$i_{\text{i.f.}} = \frac{\eta Ge}{Z_o\, h\nu}\left[\frac{2P_s\, Z_o}{\int|U_s|^2\, dA}\,\frac{2P_{LO}\, Z_o}{\int|U_{LO}|^2\, dA}\right]^{1/2} \int |U_s|\,|U_{LO}|\, dA \cos \omega_{\text{i.f.}} t$$

$$i_{\text{i.f.}} = 2\frac{\eta Ge}{h\nu}\left[\frac{P_s\, P_{LO}}{\int|U_s|^2\, dA \int|U_{LO}|^2\, dA}\right]^{1/2} \int |U_s|\,|U_{LO}|\, dA \cos \omega_{\text{i.f.}} t$$

$$i_{\text{i.f.}} = \left[2\frac{\eta Ge}{h\nu}\sqrt{P_s}\sqrt{P_{LO}} \cos \omega_{\text{i.f.}}\, t\right]\sqrt{L_M}. \tag{4-17}$$

where L_m is now defined as the mixing efficiency of the heterodyne receiver.

$$L_M = \frac{[\int|U_s|\,|U_{LO}|\, dA]^2}{\int|U_s|^2\, dA \int|U_{LO}|^2\, dA}$$

Another factor that must be taken into consideration is the tilt angle between the signal and local oscillator beams. Tilt can be generated either by misalignment of the incoming beams or by decentration of two parallel beams as shown in Figure 4-5.

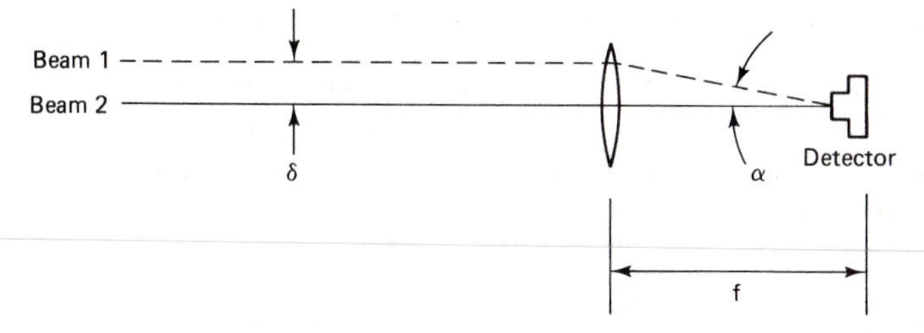

Figure 4-5 Tilt and decentration loss.

The tilt angle of α produces a loss of signal current of

$$L(\alpha) = \frac{\sin 2\pi F_s\alpha}{2\pi F_s\alpha} \quad L(\alpha) = \frac{2J_1(2\pi F_s\alpha)}{2\pi F_s\alpha},$$

rectangular detector *circular detector*

$$\alpha = \frac{\delta}{f} \text{ and } F_s \text{ is the signal focal ratio}$$

The complete geometrical mixing efficiency now becomes

$$L_m = L(\alpha)^2 \cdot \frac{[\int|U_s|\,|U_{LO}|\,dA]^2}{\int|U_s|^2\,dA \cdot \int|U_{LO}|^2\,dA}$$

Having determined an expression for $i_{\text{i.f.}}$, Equation (4-15), we can now calculate the i.f. carrier-to-noise ratio.

The shot current produced by the local oscillator is

$$i_n = (2e\,i_{dc}\,B)^{1/2} = \left(\frac{2\eta\,e^2\,GB\,P_{LO}}{h\nu}\right)^{1/2}$$

and the carrier-to-noise ratio is

$$\left(\frac{C}{N}\right)_{\text{i.f.}} = \left(\frac{i_s}{i_n}\right)^2 = 2\eta\frac{P_s}{h\nu\,B_{\text{i.f.}}}\,L_m\cos^2\omega_{\text{i.f.}}t$$

which, when time averaged (heterodyne detection) gives

$$\left(\frac{C}{N}\right)_{\text{i.f.}} = 2\eta\frac{P_s}{h\nu\,B_{\text{i.f}}}\,L_m\left(\frac{1}{2}\right) = \eta\,\frac{P_s}{h\nu\,B_{\text{i.f.}}}\,L_m$$

and when phase-locked (homodyne) detection gives

$$\left(\frac{C}{N}\right) = 2\eta\,\frac{P_s}{h\nu\,B}L_m \tag{4-18}$$

Case I, Signal and Local Oscillator Fields Matched

With the signal and local oscillator fields matched, all of the energy falling on the detector is detected in the heterodyne mode and the heterodyne efficiency increases with the size of the detector (Born and Wolf, p. 398),

$$(L_m)_I = 1 - J_0^2(x) - J_1^2(x) \qquad x = \frac{\pi r}{F\lambda}$$

where J_0 and J_1 are Bessel functions of the first kind and r is the radius of the detector, F is the f/number of the collection optics, and λ is the wavelength of the light. This efficiency factor $(L_m)_I$ is monotonically increasing function of the detector size and is equal to 0.84 for a detector whose size is the same as that of the Airy disk. (See upper curve of Figure 4-6.)

Case II, Signal Airy Function and Uniform Local Oscillator Field

This case is of particular interest because it simplifies the alignment of the signal and local oscillator beams in an actual system. The signal is a focused Airy function and the local oscillator is uniform over the surface of the detector

$$U_s(x) = \frac{2J_1(x)}{(x)} \qquad U_{LO}(x) = 1$$

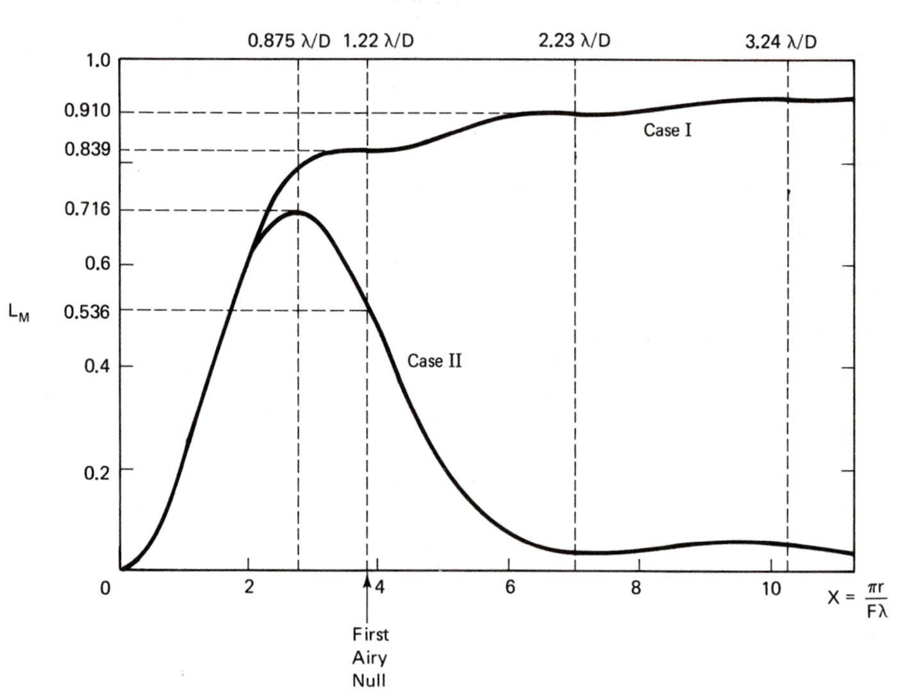

Figure 4-6 Heterodyne mixing efficiency L_M for an airy function signal distribution on a circular detector.

where the variables are the same as for Case I. The heterodyne efficiency can be evaluated and is found to be

$$(L_m)_{II} = 4\left[\frac{1 - J_0(x)}{x}\right]^2$$

This function has a peak of 0.72 corresponding to a detector radius which is 72 percent of the radius of the Airy disc. If the detector is increased in size to match the Airy disc, the efficiency drops to 0.54. (See lower curve of Figure 4-6.)

4.4.3 NEP Versus Local Oscillator Power

The condition for quantum limited operation of a heterodyne or homodyne receiver is that the shot noise produced in the detector at the i.f. frequency is sufficient to override the thermal noise in the i.f. amplifier. The shot-noise power produced by the local oscillator is

$$i_n^2 R = 2e\, i_{dc} BR = \frac{2\eta\, e^2\, GBR}{h\nu} P_{LO} \tag{4-19}$$

and the thermal noise in the i.f. amplifier is

$$4kTBN_F = 4KT_{\text{eff}}B$$

Arbitrarily large values of local oscillator power may damage the detector element. Thus, it is important to know the tradeoff between local oscillator power and receiver NEP so that an appropriate amount of local oscillator power may be specified, which is *nearly* ideal but which is within safe limits for the detector. Recalling that the i.f. current from Equation (4-17) is

$$i_{\text{i.f.}} = 2\eta \frac{e}{h\nu} \sqrt{P_s}\sqrt{P_{LO}}\sqrt{L_m} \cos \omega_{\text{i.f.}}\, t$$

and the noise current is

$$i_n = \left(\frac{4kTB}{R} + 2eB(i_s + i_d + i_{LO})\right)^{1/2}$$

then

$$\left(\frac{S}{N}\right)_{HET} = \left(\frac{i_{\text{i.f.}}}{i_n}\right)^2 = \frac{4\eta^2 e^2\, P_s P_{LO} \cos^2 \omega_{\text{i.f.}} t}{\dfrac{4kTB}{R} + 2eB(i_s + i_d + i_{LO})} \times \frac{1}{(h\nu)^2} \times L_m$$

Letting $I_{LO} >> i_d + i_s$ and factoring out $\eta(P_s/h\nu B)L_m$ yields

$$\left(\frac{S}{N}\right)_{HET} = \eta \frac{P_s}{h\nu B} L_m \left[\frac{1}{2/\eta R(h\nu/e)\,(kT/e)\, 1/P_{LO} + 1}\right]$$

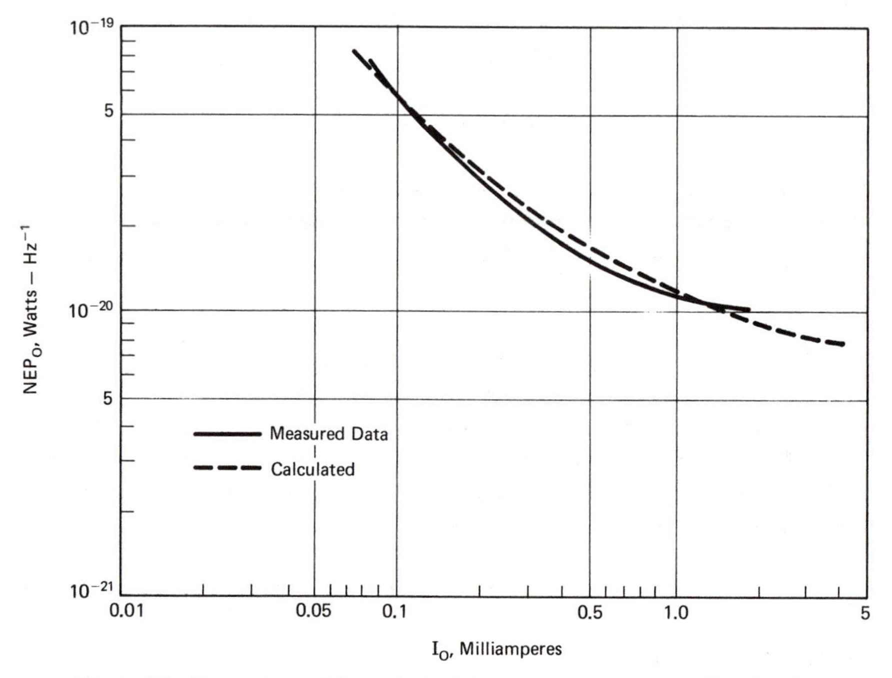

Figure 4-7 Comparison of theoretical NEP and measured data for HgCdTe PV detector at 10.6 μM.

Setting $(S/N)_{Het} = 1$ and solving for P_s yields the heterodyne NEP. The *NEP* local oscillator power dependence can then be written as

$$(NEP)_{Het} = \frac{1}{L_m}\left[\frac{h\nu}{\eta} + \frac{2kT}{P_{LO}}\left(\frac{h\nu}{\eta e}\right)^2 \frac{1}{R}\right]$$

The tradeoff is illustrated in Figure 4-7 along with actual measured values of NEP for HgCdTe at 10.6 μm. Table 4-1 lists optimum performance of selected PV detectors in the range of 3.5 to 12 μm. Both illustrations include mixing efficiency L_M, described earlier.

4.4.4 Heterodyne Conversion Gain

There are two useful definitions of heterodyne gain: conversion gain or power responsivity, and circuit gain. Conversion gain is defined as the ratio of i.f. signal power to optical signal power, electrical watts out for optical watts in.

$$\text{Conversion gain } G_1 = \frac{\text{i.f. signal power}}{\text{Optical signal power}}$$

$$G_1 = \frac{2(\eta Ge/h\nu)^2\, P_s P_L R}{P_s} = 2\left(\frac{\eta Ge}{h\nu}\right)^2 P_L R \qquad (4\text{-}20)$$

TABLE 4-1 OPTICAL RECEIVER $(NEP)_0$ VERSUS WAVELENGTH

Wavelength, Microns	*Detector Type*	η, *Percent*	$h\nu/\eta$, *W/Hz*	$(NEP)_0$, *W/Hz*
3.5	InSb	56	1.02×10^{-19}	1.72×10^{-19}
4.0	InSb	64	7.78×10^{-20}	1.18×10^{-19}
5.0	InSb	75	5.41×10^{-20}	8.20×10^{-20}
8.0	HgCdTe	39	6.38×10^{-20}	1.24×10^{-19}
9.5	HgCdTe	50	4.28×10^{-20}	7.49×10^{-20}
10.6	HgCdTe	60	3.12×10^{-20}	5.37×10^{-20}
11.0	HgCdTe	57	3.27×10^{-20}	5.47×10^{-20}
12.0	HgCdTe	30	5.55×10^{-20}	1.04×10^{-19}

$$NEP_0 = L_M^{-1}\left[\frac{h\nu}{\eta} + \frac{K(T_M + T_{IF})}{G} + \frac{h\nu}{\eta}\left(\frac{I_D}{I_0}\right)\right]$$

The magnitude of G_1 is a linear function of local oscillator power and is approximately equal to 7,500 P_L for an ideal 10 μm receiver. The effect of conversion gain is to reduce thermal noise in the receiver as can be seen by substituting G_1 in the expression for heterodyne NEP.

$$NEP = L_M^{-1}\left[\frac{h\nu}{\eta} + \frac{2kT}{P_{LO}}\left(\frac{h\nu}{\eta Ge}\right)^2\frac{1}{R}\right] = L_M^{-1}\left[\frac{h\nu}{\eta} + \frac{4kT}{G_1}\right]$$

As P_L is increased, the thermal noise is reduced to a negligible value. For a local oscillator power of 10^{-3} W, thermal noise is reduced to about 1/10 of the quantum limit.

Circuit gain is defined as the ratio of i.f. signal power to the signal power available without heterodyne conversion, or direct detection signal power

$$\text{Circuit gain } G_2 = \frac{\text{i.f. signal power}}{\text{Direct detection signal power (electrical)}}$$

$$G_2 = \frac{2(\eta Ge/h\nu)^2\, P_s P_L R}{(\eta(P_s/h\nu)Ge)^2 R} = \frac{2P_L}{P_s} \tag{4-21}$$

4.4.5 Demodulation

Demodulation of coherent carriers has been thoroughly investigated in radio and microwave communications. These techniques apply directly to optical carriers where coherent detection is used. For the present discussion, we will be concerned with three specific cases. The first of these is envelope detection of a heterodyne signal where C/N is greater than 10. The second is product detection (homodyne) where any C/N applies. The third is for band-limited FM detection where only the first sidebands of the modulated signal lie within the i.f. passband of the receiver. The C/N regime of the FM system is arbitrarily chosen to be $C/N > 10$.

Envelope Detection (Heterodyne)
$C/N > 10$

$$\left(\frac{S}{N}\right)_m = m^2 \left(\frac{C}{N}\right)_{\text{i.f.}} = \eta \frac{m^2 P_s}{h\nu B}$$

Product Detection (Homodyne)
Any C/N

$$\left(\frac{S}{N}\right)_m = 2m^2 \left(\frac{C}{N}\right)_{\text{i.f.}} = 2\eta \frac{m^2 P_s}{h\nu B}$$

$$= \frac{2\eta P_s}{h\nu B} \text{ for biphase modulation}$$

FM Detection (Heterodyne)
$C/N > 10$

$$\left(\frac{S}{N}\right)_m = 3\beta^2(1 - 0.55\beta) \left(\frac{C}{N}\right)_{\text{i.f.}} = \frac{3\beta^2(1 - 0.55\beta)\eta P_s}{2h\nu B}$$

B in the above expression is information bandwidth, and m is modulation index.

4.4.6 Frequency Tunability and Coherence Properties of Injection Laser Diodes

Coherent detection has been demonstrated using injection laser diodes and frequency-shift-keying digital modulation (FSK).[5,6] Frequency modulation has been achieved both continuously with a single laser mode[12,13,14,17] and discontinuously by mode-hopping.[11] Results have shown that in spite of the relatively noisy character of injection laser diodes, significant improvement in detection NEP over direct detection systems has been demonstrated. This section is intended to give the reader an appreciation for the difficulty in applying injection lasers to coherent systems as well as for the benefits gained.

Injection Laser Tuning

Frequency modulation or frequency shift keying is the preferred mode for coherent detection when using injection lasers. The reason is that better control of the spectrum is possible with FM as opposed to pulsed modulation.

The gain line width of a typical injection laser is nearly 100 Å wide and shifts with temperature by about +2.5 Å per degree centigrade. There are thirty or more Fabry-Perot resonator modes over the 100-Å interval. Channel substrate planar (CSP) lasers are homogeneously broadened and tend to support laser action only on one mode at a time. The Fabry-Perot cavity wavelength is tuned with temperature at a rate of +0.6 Å per degree centigrade. Simple temperature tuning of a CSP laser results in a mode hop at about every 2°C. Continuous tuning is observed over about 1 Å. However, some CSP lasers with high homogeneity will tune continuously over 6 Å or more.

Electronic tuning can be achieved by electrically changing the length of the Fabry-Perot cavity. Electronic tuning range is limited by the same factors as thermal tuning, namely by the extent of the line homogeneity. Electronic tuning may be feasible in a simple laser structure over a range of 10 Å.

The frequency shift equivalent of wavelength shift is a function of the wavelength used. At a wavelength of 13,000 Å, each angstrom of tuning is equivalent to 18 GHz of frequency tuning. At 8,300 Å, each angstrom is equivalent to 44 GHz.

Electronic tuning of the laser source can be achieved by (1) the electro-optical or Pockels effect, (2) free carrier effect, and (3) plasma dispersion. These are discussed in the following paragraphs.

1. Electro-Optic Effect

First demonstrated in 1975 by Reinhart and Logan[13] the electro-optic phase shift produced a tuning of 0.4 Å, or 15 GHz. The device was fabricated by dividing the cavity into two sections, a gain region with forward bias, and an electro-optic phase shift region with reverse bias. The active region was tapered to minimize reflections at the interface. The electro-optic phase shift region contained no active layer.

The speed of the electro-optic modulation effect is limited by the driving circuit. For all practical purposes, the EO effect is instantaneous with applied voltage.

2. Free Carrier Effect

The free carrier effect is much stronger than the electro-optic effect. The device is constructed like the EO-tuned device except that *in addition* to the EO effect, a free carrier plasma effect creates a much greater phase shift. Seumatsu[14] reported a 4 Å continuous tuning range in 1983 with a laser operating at a wavelength of 1.3 μm. The same phase shift at 0.85 μm will produce a tuning of more than 170 GHz, a factor of ten times greater than that produced by the EO effect alone.

Luis Figueroa[15] has analyzed optical phase modulation using the free carrier effect and predicts that a maximum phase shift of 3,000 wavelengths can be achieved for each centimeter of modulator length. The amount of frequency tuning thereby produced is then a function of the length of the laser cavity, modulator, and gain region. Analysis shows that as the ratio of modulator length to laser length increases, the frequency tuning range approaches 265 GHz. At 0.85 μm, this amounts to a maximum value of 6 Å.

Like the EO effect, the speed of the free carrier effect is limited only by the driving circuit. For all practical purposes, the modulation directly follows the driving voltage.

3. Plasma Dispersion Effect

Plasma dispersion is caused by the resonance phenomenon of the free carriers in the laser active region. As current is applied to the active region, the absorption edge shifts, loss is reduced, and a very large phase shift is observed. As the current reaches threshold and net gain is achieved, the phase shift stabilizes. As the laser increases in power output, the phase shift remains essentially constant.

Manning and Olshansky[16] have characterized the magnitude of the phase shift below threshold and its tuning effect on a 1.3 μm laser cavity. Scaling these data to 0.85 μm, a tuning of 440 GHz, or 10 Å is predicted. However, this work merely measured the frequency of the laser cavity below threshold.

An obvious extrapolation of Ref. 4 was conducted by Fang and Wang.[17] They constructed a laser out of a conventional cavity but which was equipped with two electrodes. One electrode was biased well above threshold and caused the device to oscillate while the other electrode was biased below threshold in the region of large phase shift. A total wavelength change of 61 Å was observed while continuous tuning of 4.1 Å was achieved.

The speed of modulation of plasma dispersion is limited by the same factors as gain modulation. In lasers of moderate power, the speed is limited to the order of 1 to 3 GHz. Experimental lasers with thin depletion layers are capable of modulation rates of 8 to 10 GHz.

Discussion of Laser Coherence

The fundamental spectral broadening mechanism for a semiconductor laser is that of random spontaneous emission events which perturb both the amplitude and the phase of the laser output. The original Shawlow-Townes theory,[18] as modified by Henry[19] and later by Vahala and Yariv,[20] results in a spectral linewidth formulation which is inversely dependent upon the laser output power P which is approximated by:

$$\Delta f = 2\omega_0 \simeq \frac{K}{P_0} \text{ Hz}, \; K \simeq 2 \times 10^4$$

The quality of the laser oscillator is reflected in the value of the constant of proportionality, K. For high-quality lasers with no extraneous noise, the value of K is 20,000. Assuming the line shape is Gaussian, the optical power spectral density can be written:

$$P(\omega) = \frac{P_o}{2\pi\omega_o} \varepsilon^{-\omega^2/2\omega_o^2}$$

where $\varepsilon = 2.718$ and the autocorrelation function can be written

$$E(\tau)^* E(0) = \frac{P_o}{2\pi} \varepsilon^{-(\omega_o\tau)^2/2} \cos \omega\tau$$

Representative sketches of optical power spectral density and autocorrelation function are shown in Figure 4-8.

The meaning of the autocorrelation function is "fringe visibility." If part of the beam is delayed for a time (τ), the fringe pattern contrast mixed with the original beam will diminish as τ increases until the two beams are totally decorrelated and the fringe pattern disappears altogether. Thus, measurement of the autocorrelation function is achieved by measuring the fringe contrast as a function of delay (τ). However, this type of measurement entails the capability of generating variable path-length delays from zero to the coherence length, which may be several hundred meters.

The inverse of fringe visibility is the phase modulation noise observed by mixing a delayed beam with itself. The measured electrical noise out of an optical mixer can be expressed as a function of the path delay (τ) or as a function of the actual path-length difference. The effect can be visualized as a homodyne detection process where the undelayed beam acts as the local oscillator and the delayed beam acts as the (noisy) phase-modulated signal. The electrical output power W from the detector consists of two terms, self noise and shot noise.

$$W = \left(\frac{\eta e}{h\nu}\right)^2 \frac{P_o^2}{4} R\, \beta^2 + 2e^2 \left(\frac{P_o}{h\nu}\right) R\, B$$

Self Noise Shot Noise

The phase modulation index β is precisely $\omega_o\tau$ or $2\pi\delta x/L$ where δx is the path

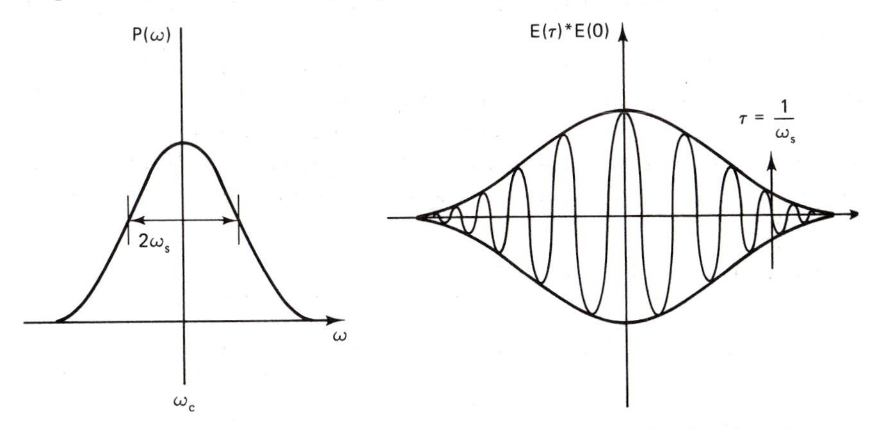

Figure 4-8 Laser power spectral density and autocorrelation function.

length difference and L is the coherence length of the laser. The self noise is low-frequency Gaussian-shaped noise with a bandwidth proportional to β. The shot-noise term is white in character such that the total shot noise power is proportional to the circuit bandwidth B. Comparing the relative magnitudes of these terms (assuming $P = 1$ mW) shows,

$$W = \underset{\text{Self Noise}}{3.1 \times 10^{-6}\, \beta^2} + \underset{\text{Shot Noise}}{1.1 \times 10^{-20}\, B}$$

The shot-noise term can be neglected for the case of interest. The self-noise expression suggests a method of measurement of the coherence length L or linewidth $2\Delta_o$. The procedure is to connect an RF power meter to the output of the detector of an interferometer. Starting with zero path length difference, gradually increase the delay length while plotting the RF power level. Figure 4-9 illustrates how reported data[21] compares with that computed from the self noise equation above. Interpolation of the data with the plot indicates that the coherence length of the laser tested is about 30 m.

A summary of measured coherence lengths is listed in Table 4-2.

4.4.7 Impact of Coherent Detection on Fiber Optical Systems

Although we treat free space communications in this book, the following brief discussion of fiber optical detection techniques is considered an appropriate aside.

Nowhere does improved detection performance have greater impact than on a fiber optical telecommunications link. Reduction of the receiver noise by 10 dB may extend the link length by 100 km. It is possible that the use of coherent detection in fiber links may eliminate the need of repeater stations in transoceanic cables! The cost impact of such a technology is astronomical.

Coherent detection in fiber optical systems is cursed with the less-than-ideal source noise properties as described in Sec. 5.6, but at the same time it is blessed with the obvious advantages of optical fiber waveguides, mixers, and coupler components. The use of these components significantly reduces the geometrical problems associated with beam optical systems as described in Sec. 5.2.

TABLE 4-2 LASER COHERENCE LENGTH *L*

Reference	Report Date	λ(m)	P1(mW)	Δf(MHz)	L(m)
Mooradian[12]	1983	0.82	10	8	37
Elsasser et al.[22]	1983	0.84	4.3	10	30
			10.0	4.0	70
Saito et al.[23]	1983	0.85	10.	3	100

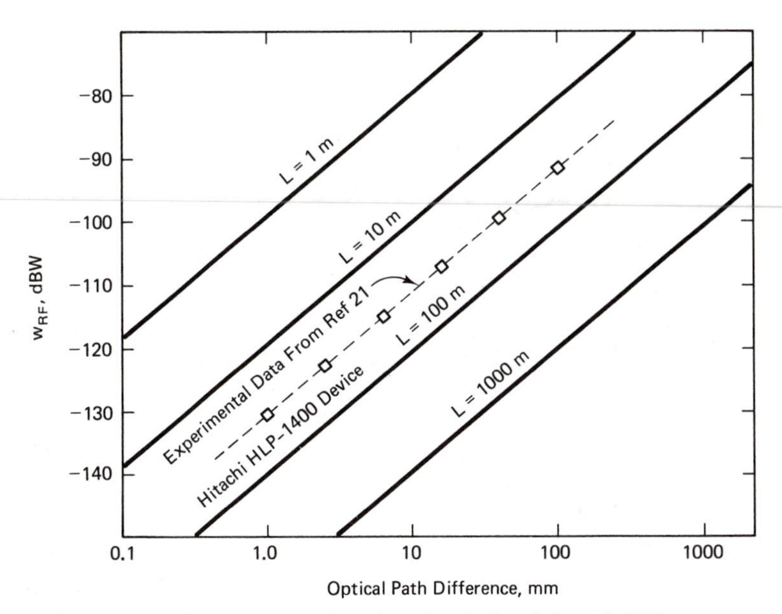

Figure 4-9 Laser self noise as a function of optical path Length Difference. (Experimental data from Ref. 21.)

The first fiber optical link system results reported by Saito[5] utilized 100 Mbps FSK and a LO laser offset frequency of 500 MHz and 1.7 GHz. The detector used was an APD with adjustable gain. Results show that the NEP was within 11 dB of the quantum limit for the 1.7 GHz offset and within 6 dB of the quantum limit for the 0.5 GHz i.f. offset. The APD gain for these results was optimized at $G = 2$. Figure 4-10 is a reproduction from Ref. 5 of the results.

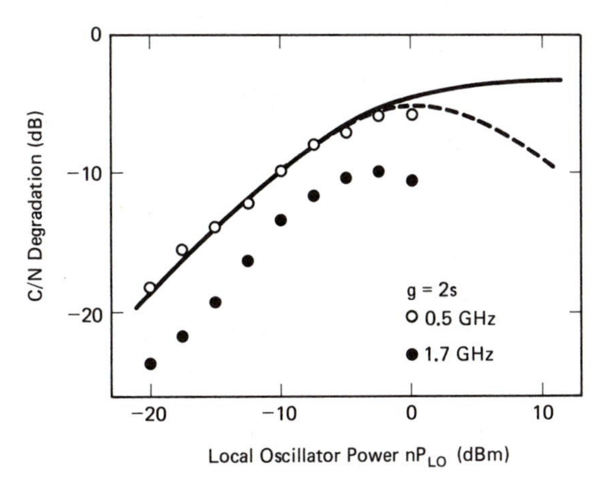

Figure 4-10 *C/N* degradation at 0.5 GHz and at 1.7 GHz for optical heterodyne receiver using injection laser diodes. 0 dB = quantum limit.[5]

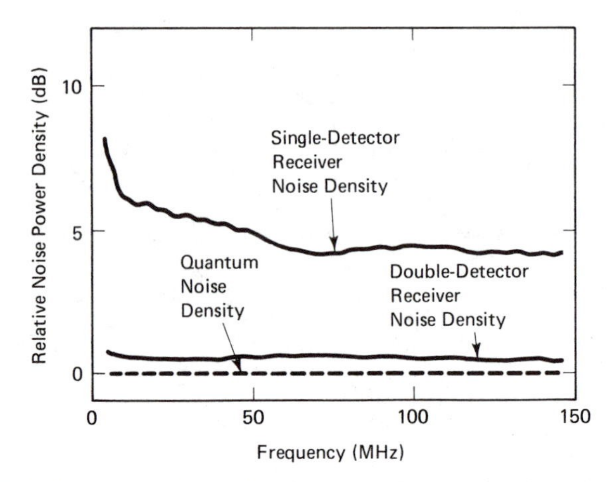

Figure 4-11 Experimental data showing suppression of excess-intensity noise with the Yuen- Chan double-detector receiver.[24]

Recently Chan and his associates have succeeded in suppressing the effects of excess-noise of the laser LO sources by the use of two detectors and a microwave hybrid coupler.[24] The technique is equivalent to the use of balanced mixers in microwave receivers. The results show a reduction in the receiver NEP from approximately 5 dB above the quantum limit for a single detector to about 0.5 dB above the quantum limit for a double detector configuration. Figure 4-11 is a reproduction of Chan's results for the single and double detector schemes.

The ultimate use of coherent detection at near infrared wavelengths both with beam optics and fiber optics is assured. It is anticipated that the techniques which are an important development in the 1980s will become an important reality in the 1990s.

REFERENCES

1. T. A. Nussmeier, F. E. Goodwin, and J. E. Zavin, "A 10.6 Micrometer Terrestrial Communication Link," *IEEE J. Quantum Electron.,* Vol. QE-10, No. 2, Feb. 1974.
2. J. H. McElroy et al., "CO-2 Laser Communication Systems for Near-Earth Space Applications," *Proc. IEEE*, Vol. 65, No. 2, Feb. 1977.
3. R. H. Kingston, "Coherent Optical Radar," *Optics News*, June 1977.
4. R. H. Kingston et al., "CO-2 Laser Radar," *IEEE J. Quantum Electron.*, Vol. QE-11, p. 308 (1975).
5. S. Saito et al., "S/N and Error Rate Evaluation for an Optical FSK-Heterodyne Detection System Using Semiconductor Lasers," *IEEE J. Quantum Electron.*, Vol. QE-19, No. 2, Feb. 1983.
6. V. W. S. Chan et al., "Heterodyne Lasercom Systems Using GaAs Lasers for

ISL Applications," *IEEE International Conference on Communications*, ICC-83, pp. 1201–07, June 19–22, 1983, Boston, MA.

7. T. G. Giallorenzi et al., "Optical Fiber Sensor Technology," *IEEE J. Quantum Electronics*, Vol. 128, p. 626, 1982.
8. A. E. Siegman, "Antenna Properties of Optical Heterodyne Detection," *Proc. IEEE*, Vol. 54, pp. 1350–56, Oct. 1966.
9. D. Fink, "Coherent Detection Signal-to-Noise," *Applied Optics*, Vol. 14, p. 689, Mar. 1975.
10. J. J. Degnan and B. J. Klein, "Optical Antenna Gain 2: Receiving Antennas," *Applied Optics*, Vol. 13, pp. 2397–401, Oct. 1974.
11. W. T. Tsang, N. A. Olsson, and R. A. Logan, "High Speed Direct Single-Frequency Modulation with Large Tuning Rate and Frequency Excursion in Cleaved-Coupled-Cavity Semiconductor Lasers," *Applied Physics Letters*, Vol. 42, No. 8, April 15, 1983.
12. M. W. Fleming and Aram Mooradian, "Spectral Characteristics of External-Cavity Controlled Semiconductor Lasers," *IEEE J. Quantum Electronics*, Vol. QE-17, No. 1, Jan. 1981.
13. F. K. Reinhart and R. A. Logan, "Integrated electro-optic Intracavity Frequency Modulation of Double-Heterostructure Injection Laser," *Applied Physics Letters*, Vol. 27, No. 10, Nov. 15, 1975.
14. Y. Tohmori, Y. Suematsu, H. Tsushima, and S. Aiai, "Wavelength Turning of GaInAsP/InP Integrated Laser with Burr-Jointed Built-in Distributed Bragg Reflector," *Electron. Letters*, Vol. 19, No. 117, August 18, 1983.
15. L. Figueroa, "Optical Phase Modulation Using Free Carrier Effect," private communication, October 17, 1983.
16. J. Manning, and R. Olshansky, "The Carrier-induced Index Change in AlGaAs and 1.3 Micrometer InGaAsP Diode Lasers," *IEEE J. Quantum Electron.*, Vol. QE-10, No. 10, Oct. 1983.
17. Z. Fang and S. Wang, "Longitudinal Mode Behavior and Tunability of Separately Pumped GaAlAs Lasers," *Applied Physics Letters*, Vol. 44, No. 1, Jan. 1984.
18. A. L. Schawlow and C. H. Townes, *Phys. Rev.* 112, 1940 (1958).
19. C. H. Henry, *IEEE J. Quantum Electron.*, Vol. 18, 259 (1983).
20. K. Vahala and A. Yariv, *IEEE J. Quantum Electron.*, Vol. 19, 1096 (1983).
21. A. Dandridge and A. B. Tveten, *Applied Physics Letters*, 39, (7), Oct. 1981.
22. Elsasser et al., "Coherence Properties of Gain- and Index-Guided Semiconductor Lasers," *IEEE J. Quantum Electron.*, Vol. QE-19, No. 6, June 1983.
23. Saito et al., "AM and FM Noise in Semiconductor Lasers," *IEEE J. Quantum Electron.*, Vol. QE-19, No. 1, Jan. 1983.
24. V. W. S. Chan et al., "Local Oscillator Excess Noise Suppression for Homodyne and Heterodyne Detection," *Optics Letters*, Vol. 8, p. 419, Aug. 1983.

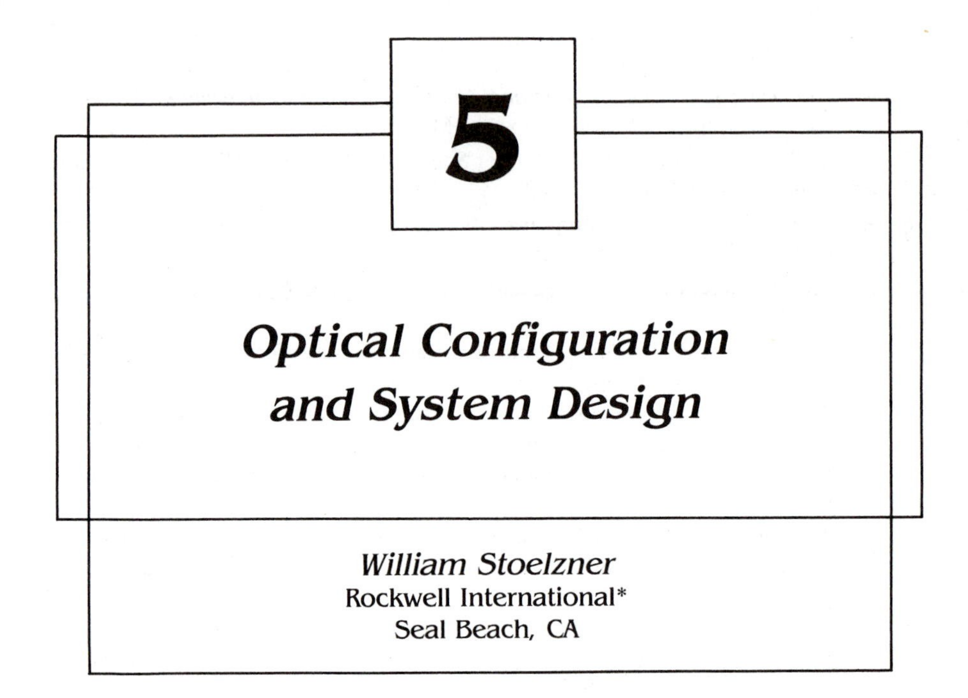

5

Optical Configuration and System Design

William Stoelzner
Rockwell International*
Seal Beach, CA

5.1 INTRODUCTION

Laser communication systems, when functioning as a satellite-to-satellite link usually referred to as the crosslink, must be designed to meet both low weight and volume, small power consumption and long life requirements common to all spaceborne instruments. These requirements in conjunction with the long range (5,000 to 50,000 nmi) commonly encountered between satellites demand a laser transmitter that is capable not only of transmitting the narrowest beam that intercepts the target accurately, but also of relaying the laser output energy to the target space efficiently. The typical laser crosslink beam spread range in the far field is about 1 to 50 μrad FWHM (full width half maximum, Sec. 5.7.1). When this fact is compared with the much wider beam spread range (500 to 3,000 μrad) commonly obtainable from laser transmitters used for terrestrial-based applications, there is little surprise that precision designs are needed not only for the internal optics, but also for the other subsystems shown in Figure 5-1. This chapter addresses all optically connected subsystems.

Many interdisciplinary technology issues and design considerations will be mentioned, but not in depth, in the ensuing discussions of the subsystem configurations and designs; nevertheless, the brief discussions of these issues serve to acquaint the readers with the options (and the limitations) of current state-of-art laser transmitters for a satellite crosslink or ranging application.

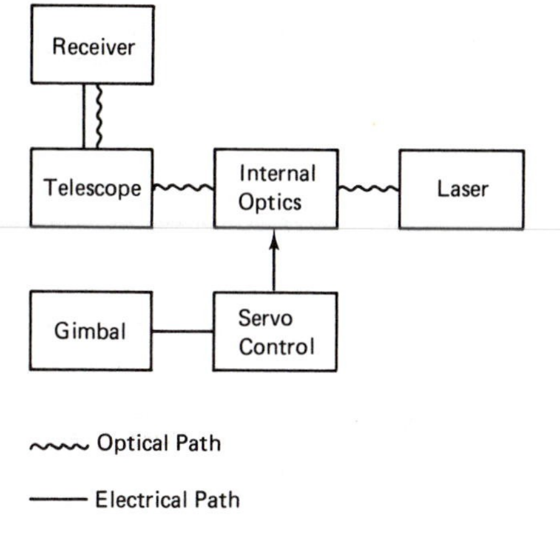

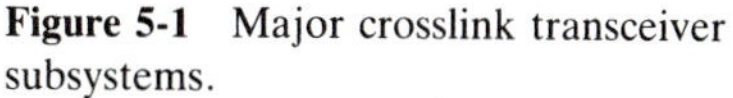
Figure 5-1 Major crosslink transceiver subsystems.

5.2 BACKGROUND

Prior to the detailed discussions of the components and methods used to synthesize and analyze the optics of a satellite laser crosslink, we will review some basic optics for clarity. More optics tutorials will be included where appropriate in the rest of the chapter. It is appropriate to review and to draw from some commonly known applications. One of the earliest formal applications of optical communication lights is the naval use of a ship-to-ship optical link—a small modulated searchlight.

The old-fashioned naval link system is simple because, beside short range and broad beam, humans point and aim the searchlight mounted on gimbals, and modulate and detect the signals. There is a one-to-one correspondence in functional blocks between the naval link and those shown in Figure 5-1. However, even trained operators limit the communication to a very slow data rate even though they are very versatile as control systems. Obviously, all such functions are performed autonomously in a satellite with greater speed and precision but with less flexibility of a human link in the functional chain.

5.2.1 Beam Spread

In this chapter, the term *beam spread* refers to the far-field pattern. The distance (Z) where far field begins for a collimated beam can be estimated as follows:

$$Z >> \frac{\pi(D^2)}{4\lambda} \tag{5-1}$$

where

Z = distance from the telescope aperture (Figure 5-2)
D = telescope aperture diameter
λ = wavelength

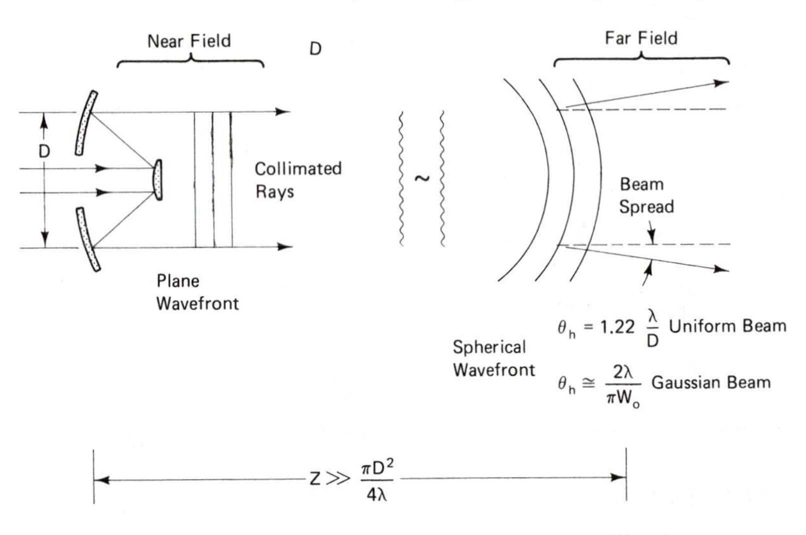

Figure 5-2 Evolution of a plane wavefront due to diffraction.

It is a consequence of the diffraction law (Sec. 5.7.2) that narrow beam spread in the far field is obtained from short wavelength, large beam diameter and nearly perfect collimation of the projected beam in the near field (region close to the telescope). For plane wavefront (collimated beam) of uniform intensity distribution and circular diameter (D), the far-field beam spread half angle (θ_h) measured to the first minimum position is:

$$\theta_h = 1.22 \frac{\lambda}{D} \tag{5-2}$$

Often, λ/D is used to estimate the beam spread of a Gaussian beam typical of a single-mode laser. This at best yields a rough estimate and can lead to very large errors unless the Gaussian beam waist radius (W_o) is only slightly less than the telescope aperture radius, and there is no obscuration (Sec. 5.8.2).

A schematic of the evolution of beam spread is shown in Figure 5-2. The fact that the nonuniform intensity distribution in the far field has a peak in the center and a zero at θ_h will be explained in Sec. 5.8.

5.2.2 Light Source

The degree of collimation of the beam is dependent strongly on two components; the first is the type of light source, and the second is the optical quality (discussed in Sec. 5.2.3).

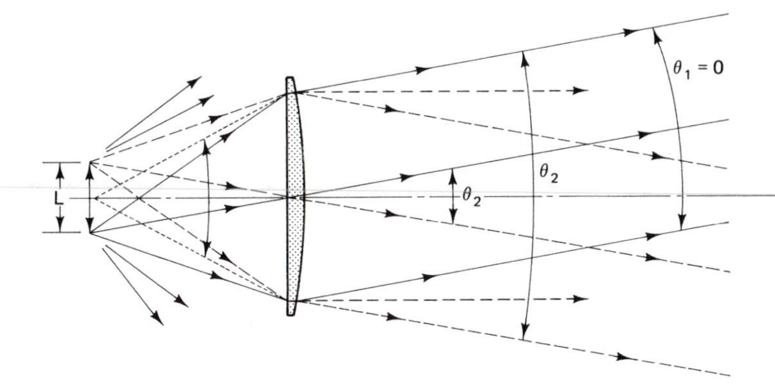

Figure 5-3 Principle of collimated beam projection.

Figure 5-3 illustrates on an intuitive level, the degree of the emergent beam collimation. The linear source L represents all light sources except one—i.e., the single-mode laser (TEMoo) output (Sec. 5.8.1)—in which case it can still be represented as a highly directional point source but not necessarily located at the same place as the linear source shown in Figure 5-3. With proper focus adjustment (Sec. 5.8.3), the shaded parallel beam ($\Theta_1 = 0$) represents what is possible with a single-mode laser. The angle Θ_2 represents the near field beam spread of common sources which are made up of many independent light sources. These emit wavefronts of many orientations causing a broader beam spread in the far field and less energy density. Multimode lasers behave in a similar manner.

One could improve the emergent beam collimation by improving the spatial coherence of the extended light source. This is done by inserting a pinhole of a diameter roughly equal to the wavelength involved; however, the price one pays is a drastic reduction in output power.

5.2.3 Optical Quality and Aberrations

High optical quality means near or total elimination of aberrations. Aberrations are the departures of the actual wavefront from those predicted by first order (paraxial) theory.[1] In this application, we can visualize aberration as any undesirable variation of the wavefront radius in a small region or over its entire surface (Figure 5-4). Aberrations must be minimized through the optical design process in order to maximize the energy density in the far field.

Small amounts of residual aberrations due to thermal or mechanical stress can be handled via methods shown in Sec. 5.8.2. Their effects can also

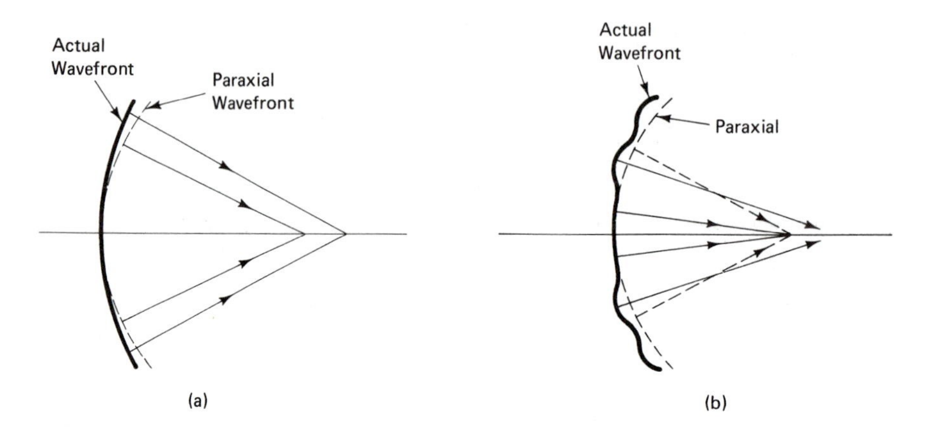

Figure 5-4a Illustrating spherical aberration in wavefront.

Figure 5-4b Illustrating wavefront aberration of more random type.

influence the transmitter design—e.g., should one gimbal the telescope or the mirror which is placed in front of it.

5.3 GIMBALS IN TRANSCEIVER DESIGN

Two-axis gimballed implementation is often necessary to perform slewing motion and coarse pointing of the laser transmitter to an accuracy range usually in the neighborhood of 5.0 to 50.0 μrad. The relative position at which it is integrated with the transmitter telescope, or antenna, affects greatly the internal optics subsystem configuration, as well as the electronic, mechanical, and servo-control designs. Therefore, careful selection of gimbal option from among the three major candidates at the initial design stage is a very cost-effective practice.

5.3.1 Gimballed Flat

The gimballed flat mirror in front of the telescope allows for a smaller load-mass on the gimbal structure, thus permitting both a light gimbal frame and a light boom truss structure. It has the additional advantage of not having to use slip rings for electrical functions of the inner gimbal and outer gimbal; an example is shown in Figure 5-5. Although the gimballed flat approach was favored by many early designs for crosslink transmitters, the other two candidates are fast gaining the system designer's attention because of the following gimballed flat weaknesses:

1. The gimballed mirror and its frame are frequently subjected to large temperature swings as a result of varying exposure to the sun. Even a mirror warping as minute as a fraction of a wavelength introduces wavefront aberrations which will cause a significant antenna gain reduction (Sec. 5.7.2).
2. The flat being relatively exposed is unprotected from impingement of thruster exhaust products. The resulting contamination reduces the optical reflection.
3. The over-sized flat mirror is often heavier than the primary mirror of the telescope, thus the actual total weight is often greater than the gimballed telescope.

5.3.2 Gimballed Telescope

With the recent advances in technology of titanium casting of the gimbal yoke, high strength but low weight gimbals are prime candidates for gimballing telescopes with apertures as large as 12 to 15 in. Another advance is the long life (> 3 years) slip rings for electrical contacts required for accurate controls of the gimbals. These technological advances have removed most major objections of gimballing the telescope except for the costs and design complexity.

The internal optical subsystem (Sec. 5.6) must be designed to propagate the beam properly along the gimbal yoke and rotation axis in order to avoid

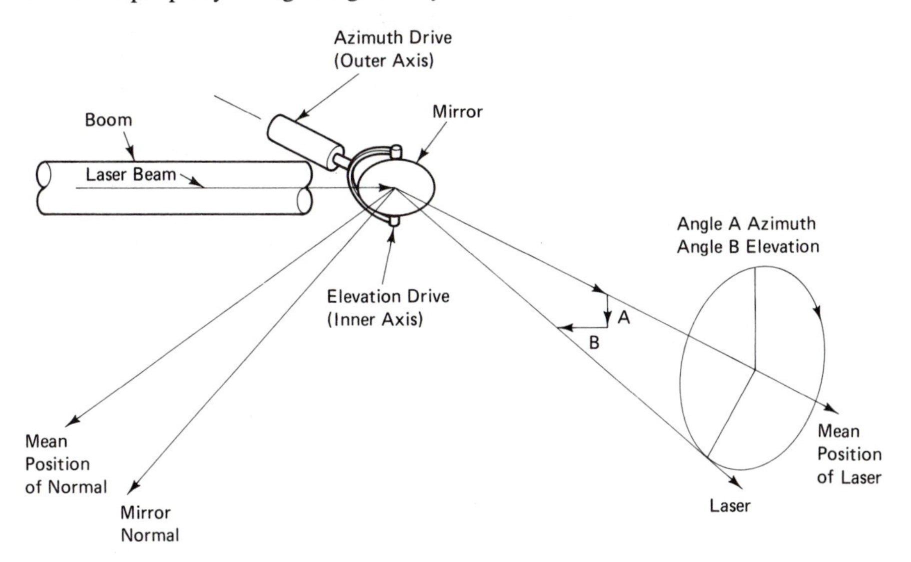

Figure 5-5 Gimballed flat beam steering.

serious vignetting and defocus problems as a result of gimbal motion. A typical and time-honored beam routing method through the gimbals is shown schematically in Figure 5-6a. This optical arrangement permits a major portion of the transmitter system weight—i.e., laser, power supply, and internal optics—to be located on the spacecraft instead of on the gimbals. For example, a 27-pound gimbal (including all the auxiliaries) will support an 8-in. telescope weighing about 11 lb; the balance of the transmitter weight which can be over a hundred pounds is off the gimbals and on the spacecraft.

5.3.3 Gimballed Platform

For a laser antenna or telescope aperture size in the neighborhood of 1 m or larger, gimballing the entire platform upon which the transmitter and receiver will be mounted is the favored approach. An excellent example of this method shown in Figure 5-6b is the NASA's Instrument Pointing System (IPS), currently being built by Dornier Systems. Its equipment platform, which is 2 m in diameter, is capable of aiming a 440-lb to 7-ton payload to an accuracy exceeding that of a much smaller gimballed telescope.

The principles and methods used in accomplishing this accuracy are illustrative of the principles of fine-pointing mechanisms located in the internal optical subsystem; this is explained in Sec. 5.6. This gimballed platform is capable of aiming a telescope within 5 μrad of a desired direction, and maintaining its accuracy when based on a spacecraft as unstable as the Shuttle. To nullify the effects of the base disturbances on the platform, the linear actuators are driven by computers which receive attitude and vibration inputs from optical and inertial sensors. For instance, the star sensor and gyroscope are instrumented to measure the gimballed platform attitude continuously, in addition to the quick measurements of changes in that attitude obtained by the accelerometers. Upon obtaining all these inputs, the computers provide instructions to the actuators for almost instantaneous changes in aiming the telescope to compensate for the undesired base motions. The complexity in the control system is severe; as such, the IPS required 200,000 software statements, whereas a typical complex satellite needed about 6,000.[3]

5.4 RECEIVER POSITION AND OPTICS

Although transmitter optics design represents the greater challenge in this application, it is not advisable to ignore the position of the receiver antenna or telescope with respect to that of the transmitter. The receiver can either utilize an independent antenna or share an antenna with the transmitter. For the case where a common antenna is used, a major concern arises in providing optical isolation between the transmitted and received beams. Furthermore, in narrow beam communications such as the crosslink, the receiver can be

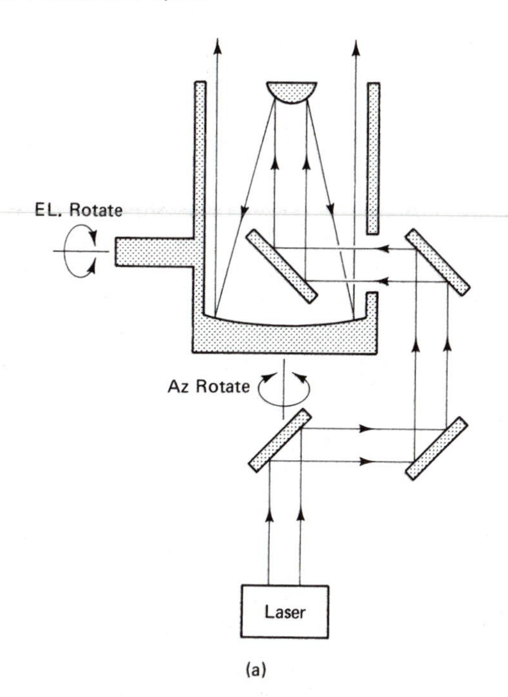

Figure 5-6a Schematic of a gimballed platform for a large telescope.

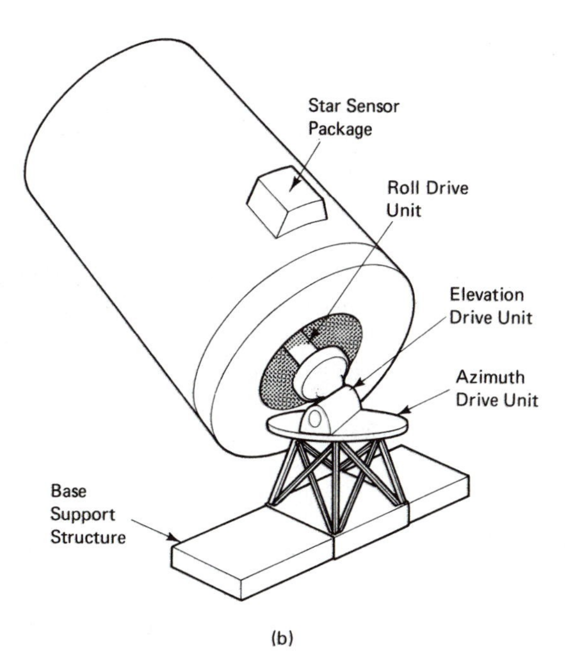

Figure 5-6b Gimballed telescope shown with the properly designed beam path.

viewed as a subsystem of the transmitter because it assists in accurate pointing by tracking the other terminal as a beacon in a closed-loop operation.

5.4.1 Shared Telescope Approach

Generally, the antenna or telescope is shared by the transmitter and receiver for savings of weight and volume and for ease of alignment. This so-called transceiver approach requires means in the internal optics to isolate the backscattered photons as much as possible from the receiver detectors. An alternate method consists of electronically blanking or gating out the receiver during the laser transmitter pulse. This sets a practical limit on data rates—i.e., <0.1 Mbit/s. Higher data-rate links require continuous optical isolation methods.

Spectral Isolation

Use of a combination of dichroic and notch filters made of quarter-wave dielectric stacks provides a receiver isolation of 10^{-9} or better, depending on manufacturer, temperature range, field of view angle, and wavelengths. However, as we shall see, there are not enough laser candidates available to satisfy all of the three following requirements to make this option very practical. These requirements are power, pulse rate, and the desired spectral separation of the incoming and outgoing beams.

Spatial Separation

For a monochromatic crosslink, the incoming and outgoing beam path behind the shared telescope often could be separated when the telescope area is large enough to accommodate the two beams (Figure 5-7).

Polarization Isolation

Another elegant monochromatic receiver isolation method which is both effective and efficient is to design the two beams with orthogonal polarizations, be they linear or circular polarized. This approach presumes that the laser output has stable polarization as in the case of Nd:YAG laser. Figure

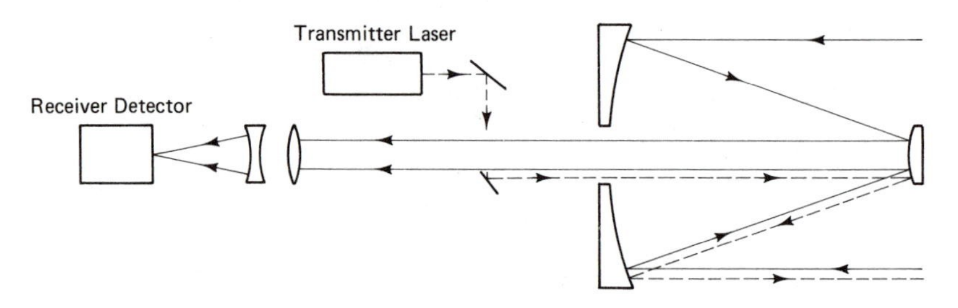

Figure 5-7 Spatial separation of transmitter and receiver while sharing telescope.

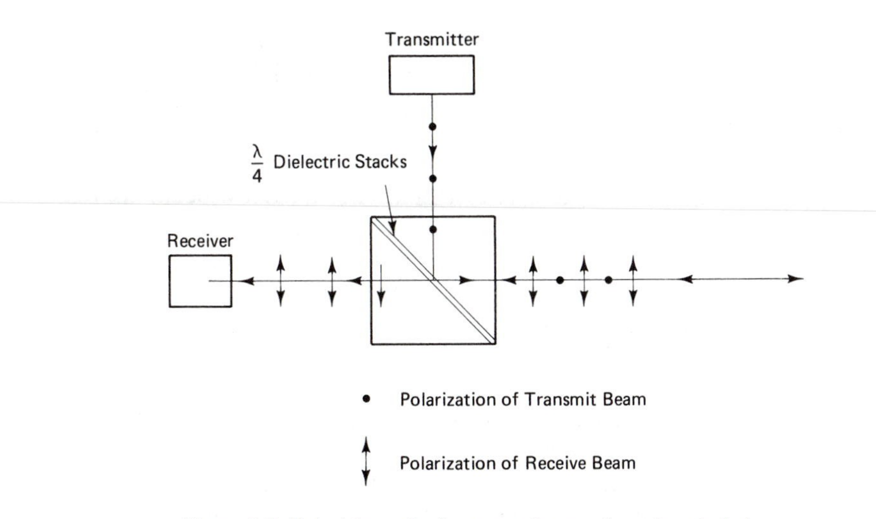

Figure 5-8 Polarizing cube for transmitter and receiver isolation.

5-8 is an example of isolating the incoming and outgoing beams via a polarizing beam splitter that is made of quarter-wave dielectric stacks. A birefringent crystal can be used instead; however, the two beams will not be at right angles. In either case, the receiver can be isolated from the transmitter by as much as a factor of 1×10^{-7}.

5.4.2 Dual Telescope Approach

There are situations where the backscattered photons can still be a major source of noise at the receiver while using the various isolation techniques described, as in the case of highly sensitive receiver or moderately powerful laser (>0.01 j/pulse). Therefore, to obtain complete optical isolation, separate but boresighted receiver and transmitter telescopes and internal optics are necessary.

The price for complete isolation is the added design complexity and mechanisms for accurate boresighting or alignment of two independent lines-of-sight. This is critical because in crosslink or ranging applications, the receiver in conjunction with control loops points the transmitter with the precision commensurate with the narrow beam.

Pointing Via Tracking and Boresighting

Although beam control is covered in Chapter 6, a brief description is given here. Consider the simple case when the target (opposite terminal) direction is known precisely and the transmitted beam has sufficient intensity within its beam spread ($\pm\theta$) to close the link. Then it is obvious that the transmitter's allowed LOS jitter angle must always be a little less than θ for consistent target illumination (Figure 5-9). However, in practice, the allowed rms LOS jitter angle or pointing jitter error is a small fraction (10 to 30 percent) of the specified beam spread (φ) (FWHM or FWe^{-2}; Sec. 5.8.1) for the following reasons:

1. The far field diffracted intensity distribution of a narrow beam approximates either a Gaussian or Bessel function, where only a small region ($2\ \theta$) around the center has sufficient intensity to be useful.
2. The beam spread angle (φ) specified is inevitably larger than the useful region ($2\ \theta$) but the specifying convention is set by those associated with diffraction theory (Sec. 5.8.1).

The more complicated case is also the more realistic case where the target direction is not known precisely. The uncertainties in receiver tracking and/or boresighting are major causes of transmitter pointing error because of the principle of reversibility of light. Another major uncertainty originates from the ubiquitous pointing jitters from control loops and beam steerer noises. The net result is that the total of tracking, beam jitter, and LOS boresighting errors has to be much smaller than the useful beam spread ($2\ \theta$). Thus we are back to where we started—i.e., calibrating the LOS of separate telescopes accurately.

Thermal and Mechanical Solution

With careful mechanical design and thermal isolators and heaters to prevent stress or hot dogging of the boresighted telescopes in orbit, the error between the two LOSs is estimated to be about 20 to 40 μrad. This presumes good alignment calibration on the ground. Naturally, this solution is limited to cases involving broader beam spread of about 200 μrad (FWHM) in the far field.

Auxiliary Focal Plane Solution

When the beam spread (FWHM) narrows to about 10 to 30 μrad, the corresponding total pointing error budget is about 2 to 5 μrad. This means the expected boresight uncertainty should be much smaller than 2 to 5 μrad.

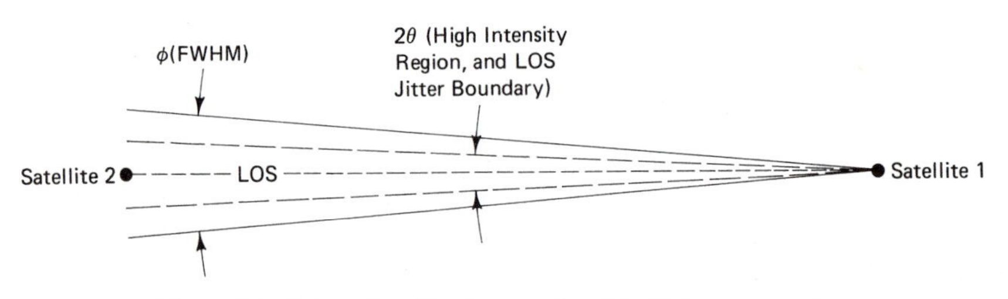

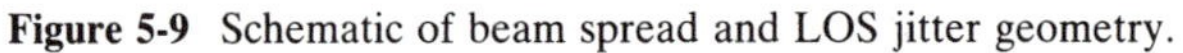

Figure 5-9 Schematic of beam spread and LOS jitter geometry.

Such a situation not only calls for good thermal mechanical designs but also for frequent on-orbit star or beacon sighting to calibrate the LOS of both telescopes: therefore, an auxiliary focal plane must be utilized in the transmitter to detect the star or beacon position. The detector signal is also the feedback to the control loops that drive beam-steering mirrors that align the transmitter LOS to that of the receiver (see Sec. 5.6.5).

Inertial Sensor Solution

As the beam spread (FWHM) narrows to about 1 to 3 μrad, commonly encountered in short wavelength laser (e.g., excimer laser) or extremely long-range communication, a more complex solution is needed. The principle for this solution is very similar to that of IPS discussed in Sec. 5.3.3. For this situation, many inertial sensors and closed-loop controls must be used to stabilize and align the two LOSs to within sub-μrad accuracy needed for the very narrow beam spread.

5.4.3 Receiver Optics

Due to the small field-of-view (FOV) (< 2 mrad) generally required by a crosslink receiver, good optical design associated with nearly diffraction limited performance can be obtained fairly routinely. Unlike the transmitter, a small amount of defocus stemming from thermal or mechanical stress does not greatly effect the receiver performance because the detectors are usually oversized to enclose more than 80 percent of the point image. This image, also known as the point spread function is called the Airy disk when it is diffraction limited, has a half angular spread as described in Equation (5-2).

The design area that needs careful attention is the aperture stop and its associated problems such as vignetting which may occur in the presence of scanning mirrors and multiple relay lenses. A review of aperture stop and its associated topics is appropriate.

5.4.4 Aperture Stop and Associates

An aperture stop is a physical element—e.g., the rim of a mirror or a lens, which limits the light cone that is incident on the image plane. In an ordinary camera, an adjustable iris aperture is designed for this purpose. In more complex optics—e.g., those encountered in laser crosslink optics—identifying the aperture stop position could be difficult without ray tracing. The commonly used technique is to find the chief ray. The chief ray is an off-axis ray that physically passes through the center of the aperture stop (Figure 5-10). This chief ray is also directed at the centers of the entrance or exit pupils, although it may not actually pass through their centers physically.[2]

Entrance and Exit Pupils

The images of the aperture stop formed by all elements in front of or in back of it are known respectively as entrance and exit pupil. As such, it means physically that any ray directed to the outside of the entrance aperture will be blocked by the aperture stop although it may pass the front optical element (Figure 5-10). The reverse case is also true. That is, any ray directed to the inside of the entrance pupil will certainly pass through the apertures and reach the focal plane. Further discussion of its usefulness in association with a beam scanner will be given in Sec. 5.7.1.

Field of View (FOV)

Based on this and our application, the half FOV can be defined as the widest angle (α), that the chief ray will make with the optical axis at exit pupil or entrance pupil and still be able to intercept the detector plane (Figure 5-10). However, this does not apply to the CO_2 laser receiver when heterodyne detection is used. In that case, the effective FOV is very small ($FOV \sim 2\ \lambda/D$), which is about the Airy disk angular substance. Poor-mixing efficiency generally results from a wide FOV in the heterodyne receiver, which causes a degradation in the signal-to-noise ratio (S/N).

Vignetting

When the cone of light accepted (Figure 5-11) from an off-axis point is significantly smaller than the on-axis cone, the effect is called vignetting. This causes an undesirable energy decrease that is proportional to the off-axis angle. The effect must be minimized and becomes a severe design problem especially in conjunction with scanning mirrors, which often are an integral part of the crosslink design.

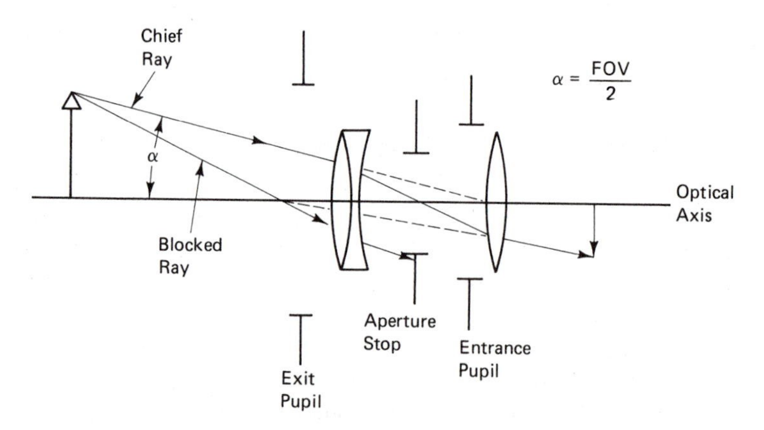

Figure 5-10 Schematic of aperture stop and associated component.

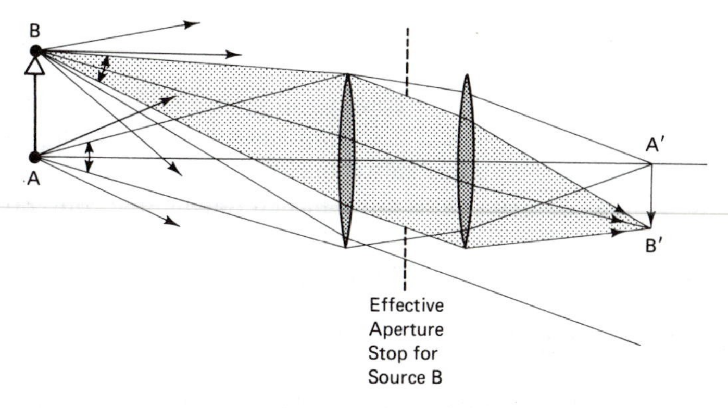

Figure 5-11 Vignetting effect (the shaded light cone represents the effects of vignetting).

5.5 LASERS

A more detailed treatment of lasers, particularly diode lasers, was given in Chapter 3. A brief review from a different perspective is given below.

For crosslink transmitter systems, the laser must have the following characteristics.

1. Light weight.
2. Sufficient power output at moderate (~1 Mbit/s) to high data rates (>100 Mbit/s).
3. Stable temporal and spatial mode, with almost mandatory circular TEMoo mode for simple and efficient optical coupling of the output.
4. Efficient modulation for pulsed or FM communication.

Unfortunately, because of all these requirements, there is not a host of lasers to vie for designers' attentions. Ignoring the excimer lasers for now because the technology is still maturing, the practical choices are limited to Nd:YAG, CO_2, and GaAlAs lasers.

5.5.1 Nd:YAG Laser

Nd:YAG is the versatile workhorse laser of industry in general and for the laser crosslink in particular. It can be operated in Q-switched, cavity dump, or mode-locked mode. As such, it can be pulse modulated from low data rate (~0.1 Mbit/s) to high data rate (~200 Mbit/s). The overall energy efficiency for Nd:YAG is low but acceptable (usually from less than 1 to 7% depending largely on the type of optical pump source—i.e., discharge lamp or diode lasers).

5.5.2 CO_2 Laser

No special transmitter design considerations are required for CO_2 laser even though it is best for coherent detection and frequency modulation. The wider beam spread as a result of diffraction of the longer wavelength does relax pointing accuracy. The more efficient heterodyne detection process in the receiver and the more efficient lasers ($\geqslant$ 10 percent) do compensate partially the effect of the wider beam spread.

5.5.3 GaAlAs Laser

This semiconductor diode laser is attractive when viewed as a single device because it requires small power, weight, and volume allocations. Furthermore, it can be modulated directly by a pulse current with excellent overall efficiency of about 10 percent.

On the negative side, there are life expectancy problems ($<$ 2 years) associated with the higher power ($\sim$50 mW aver.) diodes, and an apparent peak power limitation of about 200 mW per single device due to its small size. For typical crosslink application the diode's fan shape beam ($\sim 12 \times 60°$ FWHM) must be first reshaped into a circular beam for efficient optical coupling and then beam combined to enhance its total emitted power. Some of optical combining methods are discussed in this section since they are particular to diode laser.

5.5.4 Beam Combiner for GaAlAs Lasers

Telescope Beam Combiners

This rather cumbersome design approach is probably the only current means for combining a moderate number of diode beams with high efficiency and high optical quality. Figure 5-12 is a design of an 18-diode laser beam combiner.[4] It consists of an afocal telescope used as a beam minifier of 2 rings of diodes where the outer one has 11 and the inner has seven diodes. Before illuminating the primary mirror of the combiner, each diode output meets first a wide angle microscope objective consisting of many anamorphic lenses (Figure 5-13). Here the laser beam is transformed into a nearly circular beam without truncating much of the diode laser output.

The typical combiner performance and design data are shown in Table 5-1[4] and its dimensions are shown in Figure 5-12. A few designers had tried, but failed to reduce the output beam spread substantially without the attendant enlargement in combiner size and/or reduction in power throughput. Therefore, this subtle design difficulty could limit the combiner method to applications where the ultimate desired beam spread out of the transmitter antenna is not less than 30 μrad (FWe^{-2}). The rationale is simply that a 900

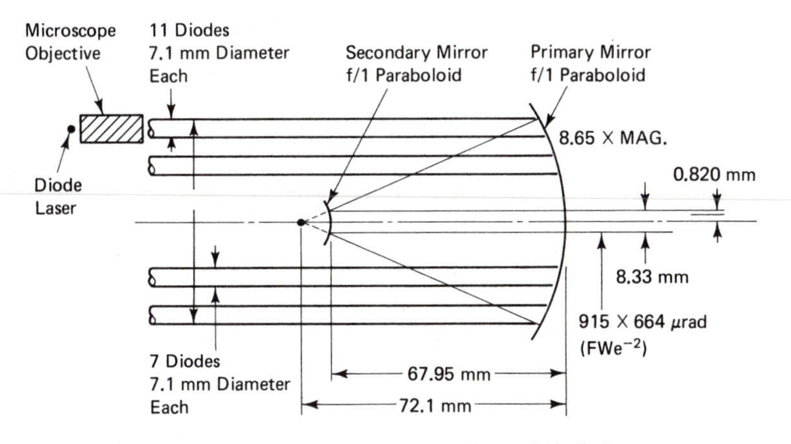

Figure 5-12 Telescope beam combiner of 18 diodes.

μrad combiner output will result in a 30-μrad transmitter beam while assuming a practical afocal magnification limit of ≤30. (Sec. 5.6) for the telescope antenna.

The strengths of the telescope combiner approach is not so much in its ability to increase dramatically the signal margin by projecting a very narrow beam (<<30 μrad) but mostly in its ability to do the following:

1. Relax the pointing accuracy requirement to around 6.0 μrad because of the broader beam (≥30 μrad), thus removing the very difficult technology issues associated with 2-μrad pointing.
2. Increase the transmitter reliability dramatically because its 18 diodes per beam would lead to graceful degradation.

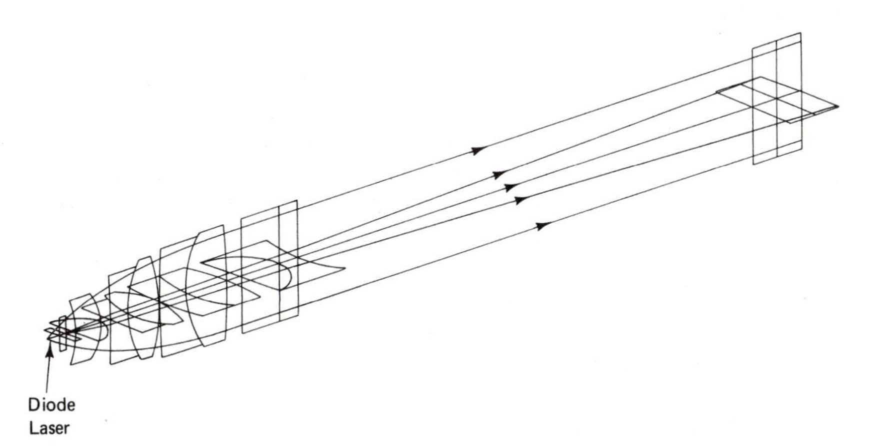

Figure 5-13 Example diode laser collimator using microscope objective (anamorphic).

TABLE 5-1 OPTIMIZED COMBINER DESIGN[4]

18-diode beams in as compact a bundle as possible	
>50% throughput	
900-μrad maximum $1/e^2$ far-field divergence	
8 to 12 mm output bundle diam	
Athermalized design	
~1-μm positioning accuracies for all degrees of freedom	
Minimum weight and size	
3.5-lb without electronics	
76 × 76 × 152 mm	
f/1 optics with λ/8 quality at 0.6328 μm (beam reducer)	
Simple spherical or anamorphhic lens arrays (collimators)	
Lens/diode alignment tolerances	
Focus adjustment	±3 μm
x-y alignment	±1 μm
Thermal expansion	≤0.5 μm
Jitter	≤0.5 μm
Angular alignment of diode	±0.1 deg

Collinear Combiner

The dichroic filter and the orthogonal polarizing cube mentioned earlier are also excellent beam combiners with good throughput and collinearity, but they are limited to two diodes. Prisms and gratings when used as beam combiners have low throughput and difficult optical layout geometry; not to mention the problem of maintaining an array of multicolored diode lasers, with proper spectral separations by temperature control of each individual diode.

5.6 ANTENNA/TELESCOPE

To project a collimated beam efficiently, the commonly used antenna is usually one of the many afocal telescopes (Figure 5-14a) with the Cassegrainian type as the most popular. Another advantage of this design is that any angular movement of the beam due to internal disturbances is demagnified by a factor $M = f_1/f_2 > 1$ upon projection (Figure 5-14b). This property minimizes the effects of pointing jitter and misalignment bias.

Telescope optical design in this section will be limited to those aspects particular to crosslink laser transmitters.

5.6.1 Beam Collimation

Laser satellite communications applications require a degree of collimation of the transmitted beam far beyond the conventional telescope design requirements. For instance, typically the degree of collimation required for

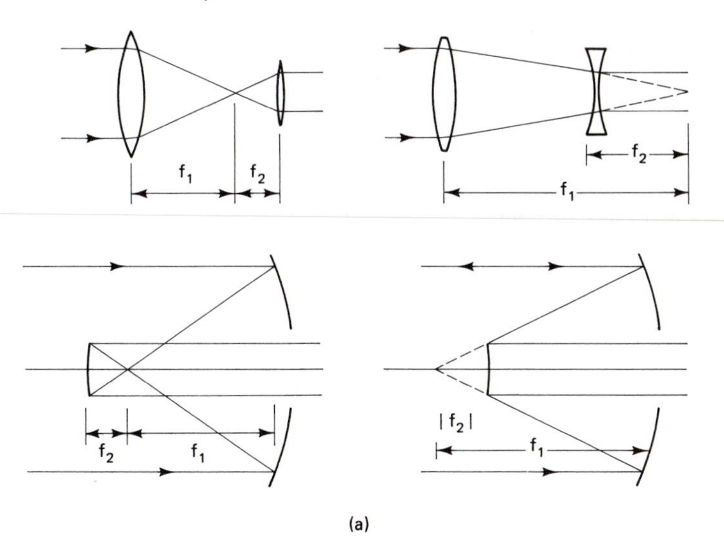

Figure 5-14a Afocal telescope arrangements (the left sketches are Gregorian and the right sketches are Cassegrainian).

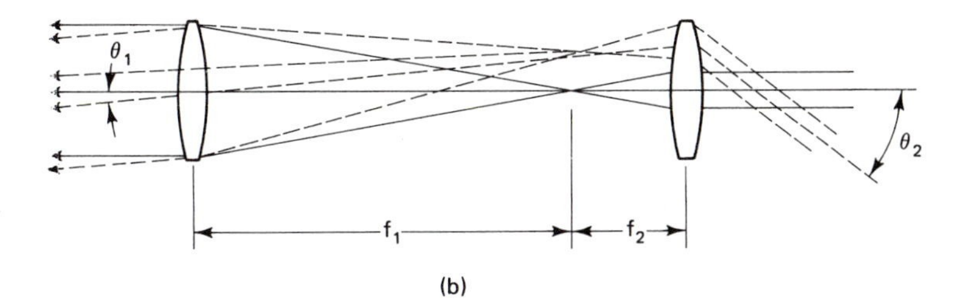

Figure 5-14b Demonstration of demagnifying the angular movement θ_2 upon projection ($\tan \theta_1 = \tan \theta_2$ (f2/f1)).

a 9.5-cm telescope aperture is about 50 to 150 km wavefront radius at the exit plane of the telescope. The reason stems from the fact that only a collimated beam or high antenna gain (Sec. 5.8.2) can minimize propagation loss of signal power in vacuum (Figure 5-15).

The most sensitive part of a well-designed telescope in causing decollimation has been found to be the spacing between the primary and secondary mirrors. The following paraxial lens equation will show that the acceptable spacing tolerance (X) is about a few microns in length assuming collimation of $R \sim 100$ km and a 0.38-m primary mirror focal length (f). That is:

$$X = f^2/(R - f) \tag{5-3}$$

The result of a more exact analysis (Sec. 5.7.3) is shown on Figure 5-16, and confirms the usefulness of Equation (5-3) for quick estimation.

In view of the critical spacing tolerance needed between the primary and secondary mirrors, the temperature of the metering rods must be controlled to within about 10°C. This means that temperature sensors and heaters often are required to maintain the desired collimation while in orbit.

5.6.2 Mirror Weight

The primary mirror weight of the telescope is often the departure point for the mechanical, thermal, optical, and servo-control design considerations. This mirror has to be rigid enough to maintain its figure to a fractional wavelength ($<\lambda/10$) while in orbit, so obviously it cannot be too thin. Table 5-2 is a quick reference to the weight range of a few actual primary mirrors based on reliable final design data.

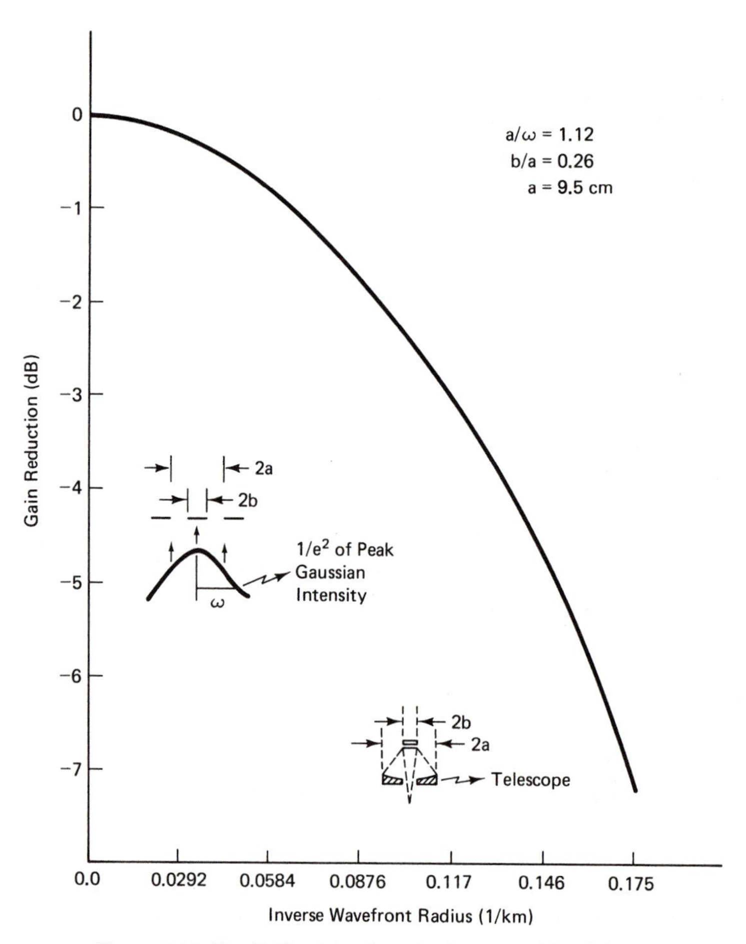

Figure 5-15 Far-field axial gain reduction caused by defocus.

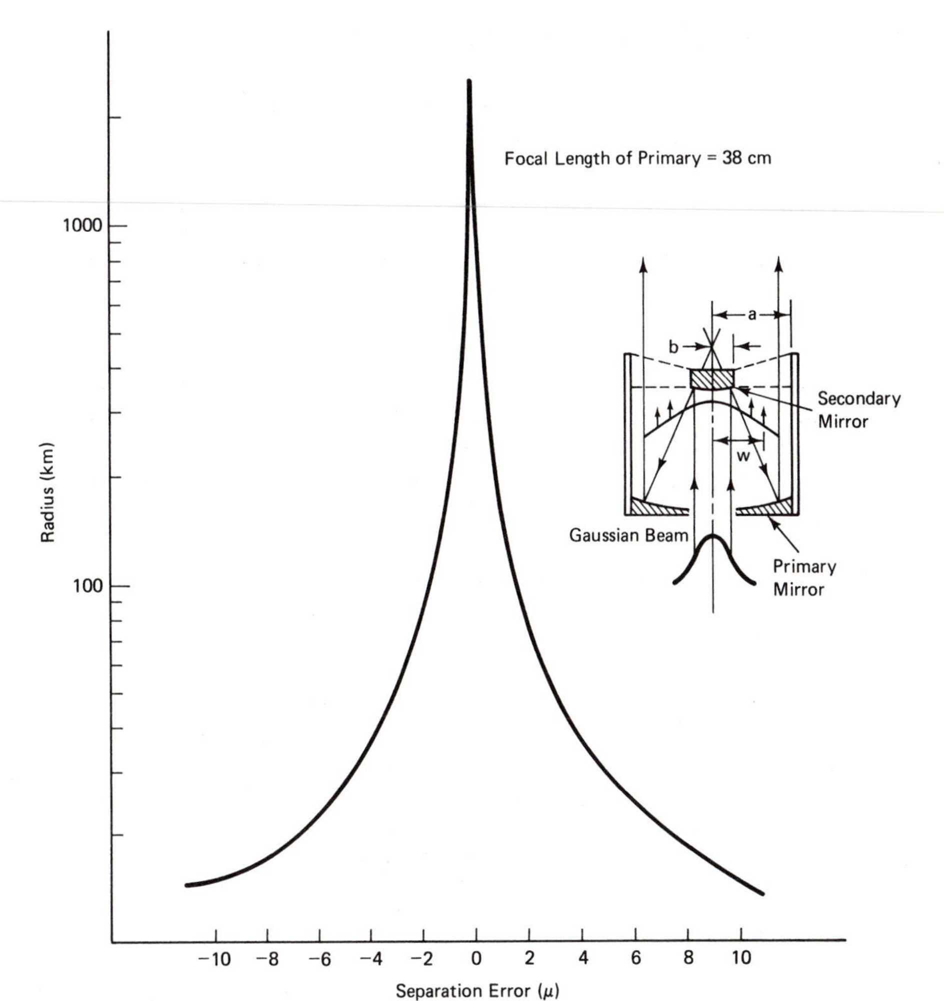

Figure 5-16 Wavefront radius variation due to separation error between primary and secondary mirrors.

TABLE 5-2 EXAMPLE OF PRIMARY MIRROR WEIGHT

Diam. (cm)	*Material*	*Wt. (lbs)*
19	ULE-Frit Bond	4.5
35	Aluminum	7.4
55	Aluminum	24.0
81	Fused Quartz-Egg Crate	90.0
99	ULE	140.0

5.7 INTERNAL OPTICAL SYSTEMS OF TRANSCEIVERS

The internal optical system (IOS) has many important functions that are vital to the success of the communication link in orbit.
These are listed as follows:

1. It must deliver to the telscope a collimated beam of proper diameter with minimal amount of beam truncation, vignetting, and optical aberration. The laser output is assumed to be in the TEMoo mode.
2. It must do the point-ahead function accurately with pointing errors and jitter much less than the useful beam spread.
3. It must be able to react quickly in restoring the transmitter LOS orientation. This also applies to receiver LOS which tracks the target image when the internal optics is a part of the transceiver (Sec. 5.7.4).
4. It must perform periodic or preferably continuous alignment or boresighting measurement of the LOS between the transmitter and the receiver while in spaceborne operation.
5. It must accommodate the selected parameters such as laser wavelength, dichroic, or polarization without conflicting with the other four functions mentioned.

To perform all these functions, candidate optical system architectures are rather complex and intricate. Some of the successful and yet simpler internal optics configurations will be discussed. We can illustrate the synthesis of a transmitter system with the options discussed. However, we will first make a brief detour into some discussion of important functions—e.g., beam steering, point-ahead steering, and other optical topics.

5.7.1 Beam Steerer

A modern beam steerer usually is made by tilting or rotating a flat mirror that alters the direction of the optic axis and the LOS of an instrument on command from the image position sensor (Figure 5-17). The closed-loop control of either type 1 or type 2 can have a bandwidth as high as 300 Hz, thus capable of quick response in stabilizing the LOS from outside disturbances—e.g., target motions, gimbal frictions, or structure vibrations. The tilting mechanism is generally a galvanometer that has a rotor shaft with a flat mirror attached. However, some tilting flat devices are driven by piezoelectric actuators. There also are options in implementing the rotating mirror, as well as in the combining arrangement to obtain the required scanning, some of which are shown in Figure 5-18a, b, c.

Regardless of mechanism, the beam steerer should best be put in a collimated region and placed as close to a relay pupil as possible to minimize vignetting on both the transmit and receive beams as the mirror is rotated.

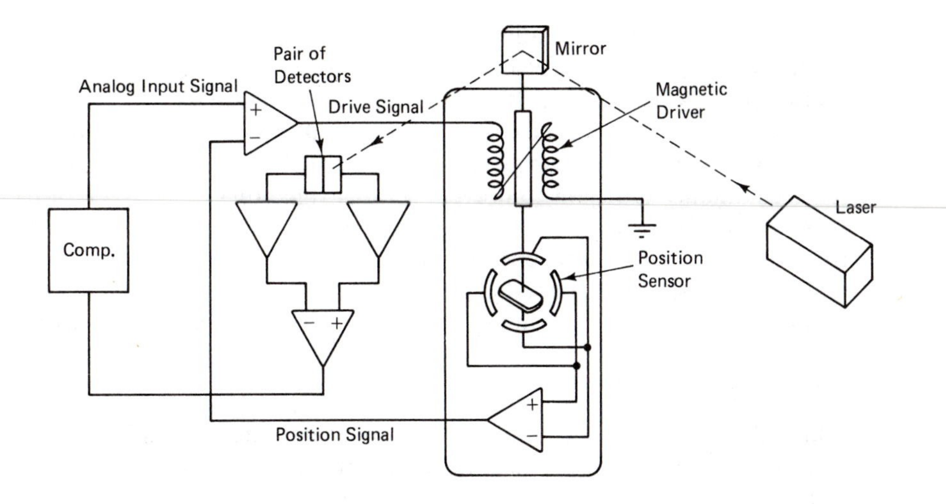

Figure 5-17 Schematic of a closed-loop controlled mirror scanner.

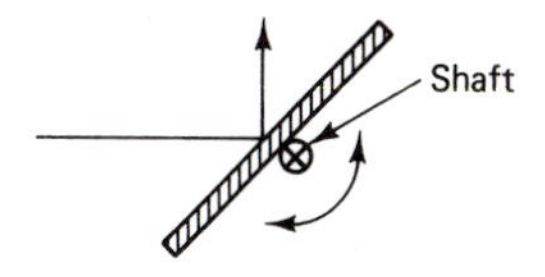

(a) Shaft to Beam Rotation Factor of 0.5

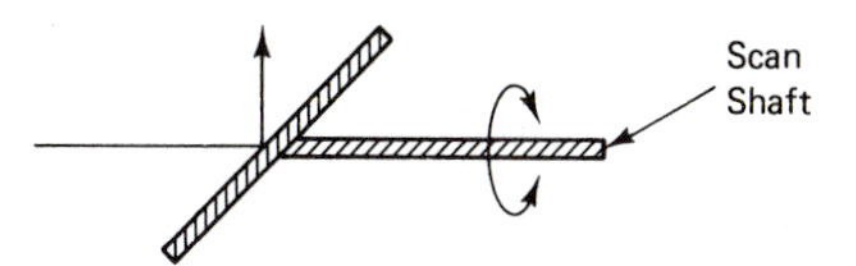

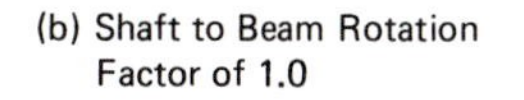

(b) Shaft to Beam Rotation Factor of 1.0

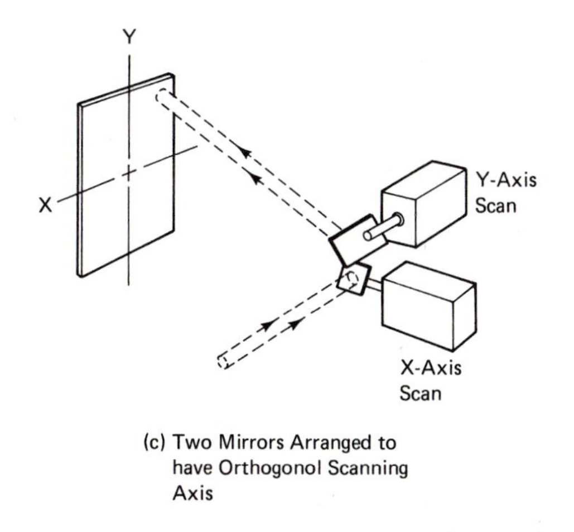

(c) Two Mirrors Arranged to have Orthogonol Scanning Axis

Figure 5-18 Mirror rotation technique for scanning.

For example, one useful technique is to image the secondary mirror of the telescope onto a plane midway between the mirrors of the beam-scanning unit, as shown before (Figure 5-18c). However, small residual beam walk on the secondary mirror is expected due to the axial separation between the scanning mirrors and their midway relay pupil. A slightly enlarged secondary mirror is commonly sufficient to prevent serious vignetting.

The scanning mirror, being the actuated member of a close-loop servo-control, will suffer bandwidth dependent angular jitters or angular noise. This jitter in the target space can be reduced in a number of ways, such as increased S/N, enlarged afocal optics magnification, and decreased servo-nonlinearity.

Point-Ahead Steerer

In order to carry out communications between satellites with relative motion, the transmitted beam must be aimed at a point in space where the receiver satellite will be at the time of arrival of the beam.

Achieving accurate point-ahead angles becomes critical because of the long-range and the very narrow beam divergence angles. The minimum point-ahead angle is determined by the expected target transverse movement (ΔX) in the roundtrip time (T) of a light pulse (Figure 5-19).

$$\delta = \frac{\Delta X}{Z} = \frac{2V}{C} \tag{5-4}$$

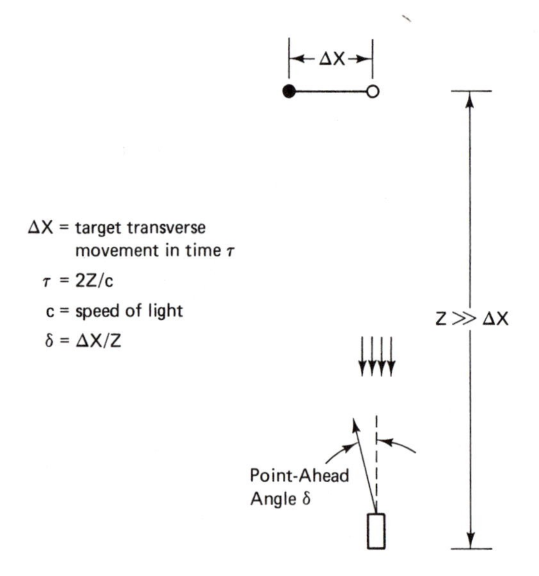

Figure 5-19 Schematic of point ahead geometry.

where

Z = range
V = target transverse velocity
C = speed of light

For example, the point-ahead can be as large as ±40 μrad in geosynchronous orbit where the range between terminals can be larger than 80,000 km.

The position of the point-ahead steering mirror unit with respect to the tracking-steering mirror should be carefully considered. For instance, the colocated tracking and point-ahead often involves a beam-steering unit that has dual functions, and consequently has more sophisticated mechanisms and control loops than the single function unit. Other design ramifications will be addressed in the sections on IOS configurations—i.e., Secs. 5.7.6, 5.7.7, and 5.7.8.

5.7.2 IOS Relay Optics

Among the many functions placed on the relay optics, such as beam shaping and pupil imaging, is the one to provide afocal magnification through stages so as not to place all the burden on the final afocal telescope to save weight. That is, the mirror weight is strongly dependent on its diameter (D) (Table 5-2), which influences magnification M as follows:

$$M = \frac{D\text{ (primary)}}{D\text{ (secondary)}} = \frac{f_1}{f_2} \tag{5-2}$$

Figure 5-20 illustrates the magnification allocations within the IOS of a lightweight transmitter design.

5.7.3 Optical Material in Space Radiation

Utilizing refractive elements for relay optics is not only a more convenient design approach but is optimum for the weight and volume considerations. The lens materials must be chosen to withstand space radiation. An optical

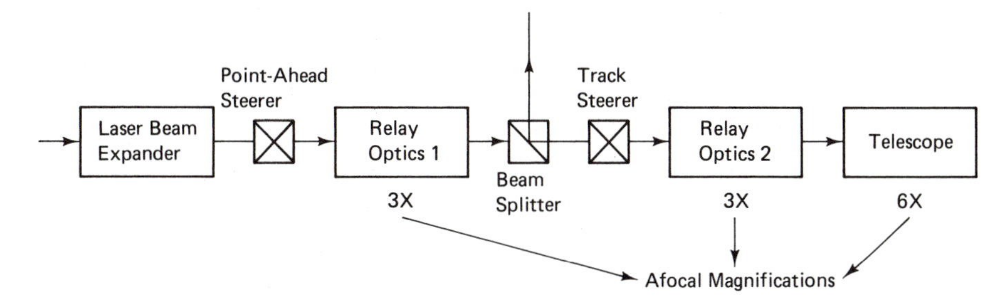

Figure 5-20 Example of allocating afocal magnifications.

material that has the least amount of impurity generally suffers the least transmittance loss when subjected to a prolonged radiation dose.

The common lens materials good for spaceborne instruments are fused silica and BK-7 glass. Their transmission extends from the near UV to the

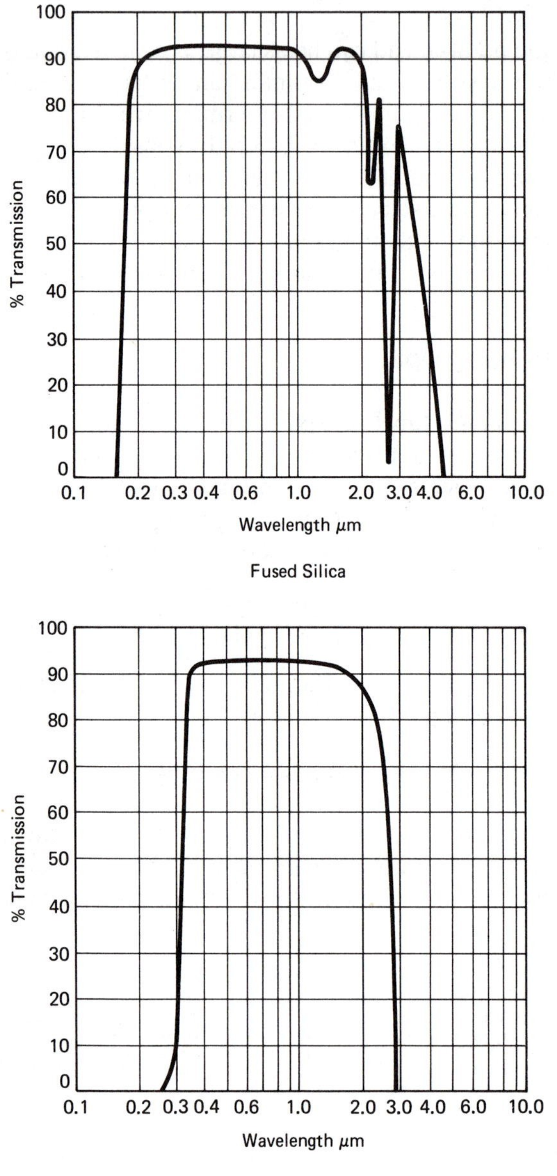

Figure 5-21 Transmission of optical materials.

near IR spectral region (Figure 5-21). Fused silica of good quality has been tested to determine its radiation induced transmittance change. The measured decrease is about 0.3 percent after a dose of $\sim 1 \times 10^6$ rads. Should the fused silica be used as an external window where a higher total dose is expected ($>10^{10}$ rads), then the estimated transmittance loss is about 1 percent.

Although the internal refractive optics require one or two good radiation resistance materials because of the monochromatic laser, this situation does not apply to the spectral filter located in front of the detectors. A narrow-band spectral filter often contains many proprietary materials, stacked in layers of quarter-wavelength thickness that may not be radiation resistant. Large losses (10 to 30 percent) can be expected from spectral filters unless their constituent materials are specifically selected for spaceborne operations.

5.7.4 Optical Transmittance

Total transmittance of the laser transceiver varies naturally from case to case, depending on wavelength, materials, optical coatings, and total number of optical elements. Here, we shall briefly indicate what can be expected within the state-of-art in the visible to near-infrared spectral region.

With good optical coatings, each refractive or reflective element (not including filters) can be expected to have transmittance or reflectance exceeding 0.99. The spectral filter transmittance, however, is lower, typically 0.5 to 0.7 for 10 to 20 Å wide filters and 0.8 to 0.95 for wider band filters. These filters can be very sensitive to temperature and alignment changes—therefore, these effects must be accounted for in the design.

For a crosslink transceiver containing twenty to thirty coated optical elements including spectral filters, the net transmittance expected is about 20 to 50 percent.

5.7.5 Air to Vacuum Focus Shift

Adjustment must be provided in the optical design to allow for focus shift in the transmitted beam when testing the beam in the laboratory, unless a large vacuum chamber is available. As mentioned in Sec. 5.6.1, focus or collimation is a very critical parameter of the transmitter performance.

The amount of defocus or decollimation expected can be as much as $\lambda/3$ to $\lambda/6$ rms, in a transceiver which has 10 to 20 refractive elements in the path. As we shall see later (Sec. 5.8.2), this must be corrected in the design. The cause of decollimation is the differences in refractive index between the material and vacuum and air, which for fused silica are, respectively, 1.450021 and 1.44963.

One solution is to make certain that the spacing between the selected optical elements is accurately adjustable and predictable. This means that very careful design consideration is necessary.

5.7.6 IOS Configuration 1

Figure 5-22 is an illustration of a beam steerer unit capable of doing both tracking and point ahead. The Risley prisms[5] compensate for the bias, caused by the point-ahead mirror, by simply steering the received signal onto the center of a quadrant detector; this results in nearly real time alignment of the two LOSs. Periodically the shutter will open so the retroreflected residual laser beam is also centered on the quadrant detector thus establishing the reference point for Risley offset. The Risley prism can be replaced by a pair of scanning mirrors as the beam steerer, but a fast response often is not necessary. Obviously, in view of Figure 5-22 the architecture for dichroic IOS becomes very complex since in addition it is required to meet the redundancy and wavelength interchangeability requirements that often have been designed into the space communication system.

5.7.7 IOS Configuration 2

A good technique when using the simpler one-function beam steering unit is illustrated in Figure 5-23. Placing the point ahead unit in the optical path not shared by the receiver provides the advantage of eliminating the beam steerer for offsetting the receiver axis.

When one compares the two IOS configurations shown, the optical unit that allows one to obtain near continuous real time alignment between the LOS of the transmitter and receiver is what may be called the "antiparallel

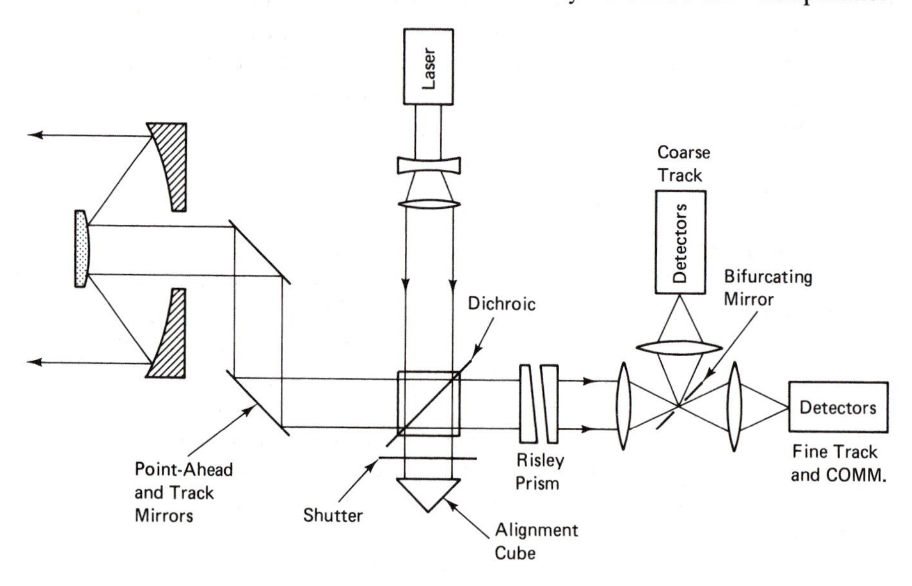

Figure 5-22 Internal optical subsystem configuration 1 (omitting most of relay optics).

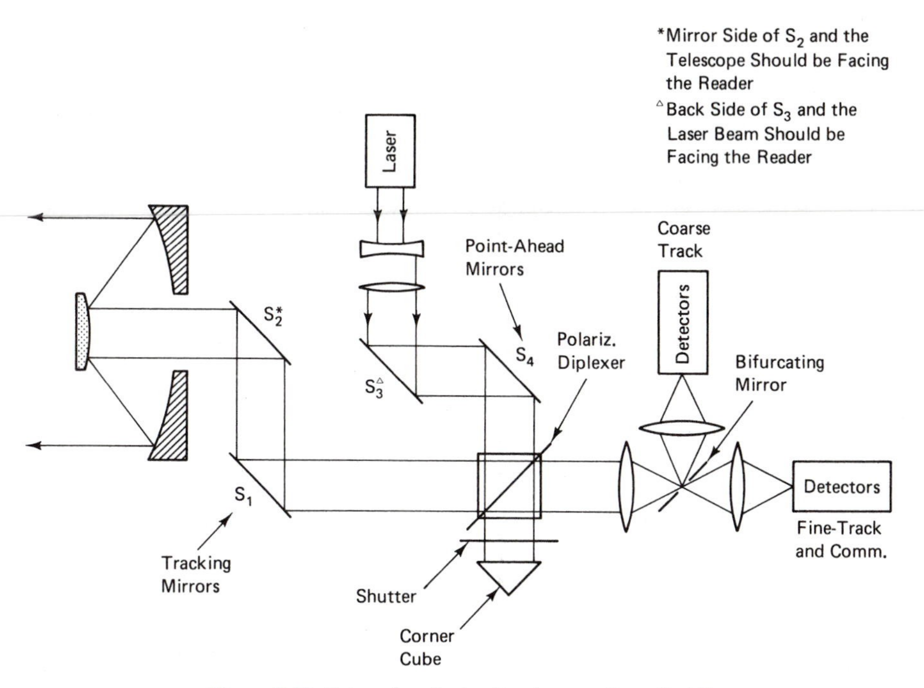

Figure 5-23 Internal optical subsystem configuration 2.

beam splitter." This unit consists of a diplexer (dichroic or polarizing), which reflects almost all the incident laser energy, and a retroreflector (corner cube), which causes the minute laser energy being leaked by the diplexer to propagate in a final direction exactly opposite to the dominant emitted beam. No matter how the incident angle may change, this unit produces two antiparallel beams (Figure 5-24). Obviously, the IOS architecture will become even more complex should one overlook the value of this antiparallel beam splitter unit.

5.7.8 IOS Configuration 3

As was discussed in Sec. 5.3.2, a separate receiver telescope boresighted with the transmitter still could impact transmitter IOS designs mainly because of the addition of an auxiliary detector plane in the transmitter.

A design for a dual telescope approach is shown in Figure 5-25. In this example, the transmitter IOS is essentially that of configuration 2; however, the detector plane is no longer required to detect received pulses at a high rate typical of a communication channel. However, it still has to do the following for the transmitter in conjunction with an added small computer that coordinates the detectors in either telescope:

1. Provide a signal to the beam steerer to calibrate the transmitter LOS

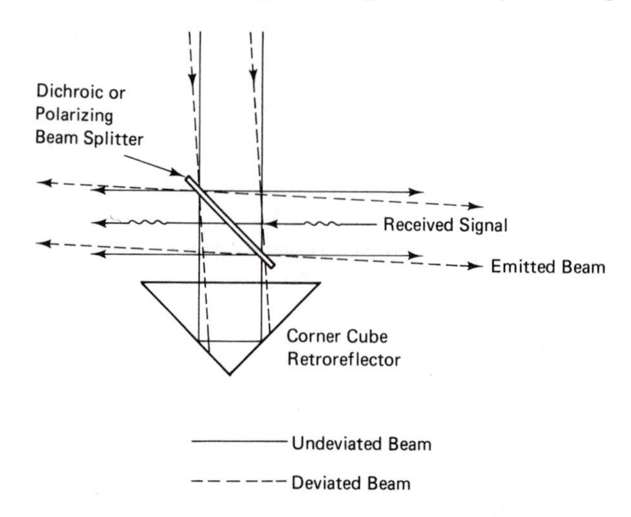

Figure 5-24 Antiparallel beam splitter for alignment.

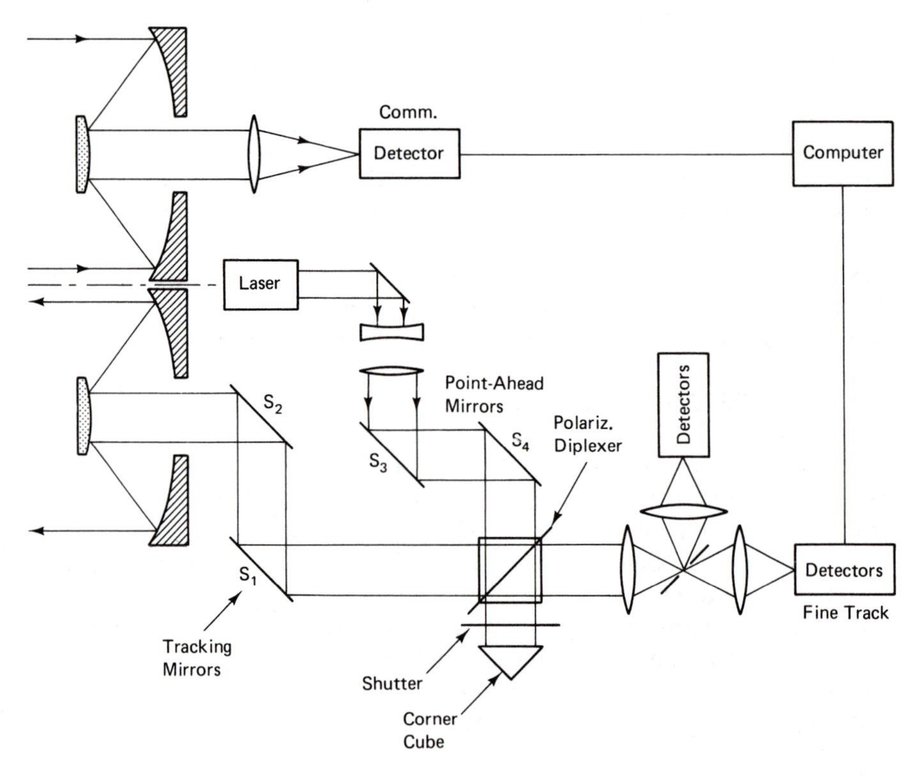

Figure 5-25 Internal optical subsystem configuration 3.

with respect to that of the receiver by occasionally sighting a star or a beacon.
2. Measure the point ahead angle of the beam.
3. Provide feedback signal to the tracking mirror by detecting and integrating the received signal.

All three configurations discussed could use the following design tip. Means should be provided to diffuse or to deflect the small but annoying amount of laser beam reflected back from the vertex region of the secondary mirror into the transmitter and its focal plane; otherwise, the budgeted isolation factor can turn out to be inadequate.

5.8 OPTICAL TRANSMITTER ANALYSIS

The efficient laser crosslink transmitter is required to deliver a narrow beam that is limited only by the far-field diffraction. This means that a complete optical design must be done by methods grounded in geometrical and wave optics. Modern approach begins the design with a rough layout drawn from the designer's experience and knowledge, but finishes it in steps by using modern computer optimization programs. As useful as these codes are, they may not be designed to minimize beam truncation and beam divergence of a Gaussian beam. Defocus or truncation often appears in a Gaussian beam being propagated through an optical system designed only by the geometrical optics principles. Compensations can be obtained successfully by using results from a separate Gaussian beam analysis, which is outlined in the next section.

5.8.1 Gaussian Beam Properties

A brief review of Gaussian beam definitions is given before its propagation properties are discussed.

When the transverse intensity distribution of a plane or spherical wave has a bell-shaped curve that is described by the following equation,[1] it is called a Gaussian beam (Figure 5-26a)

$$I(x,y) = \frac{2}{\pi}\frac{1}{W^2}\, e^{-2(X^2+Y^2)/W^2} \tag{5-5}$$

where W = spot radius or spot size.

A point at the spot radius is 0.135 ($1/e^2$) times less intense than the maximum intensity located at the center of the beam. The percent of the total energy within a circle of radius a is shown on Figure 5-26b and is calculated using Equation (5-6):

$$\frac{\mathrm{I}(a)}{\mathrm{I(total)}} = 1 - e^{-2a^2/W^2} \tag{5-6}$$

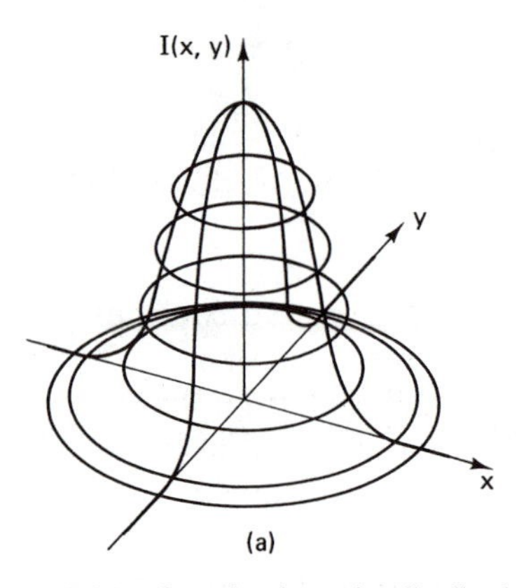

Figure 5-26a Gaussian intensity distribution.

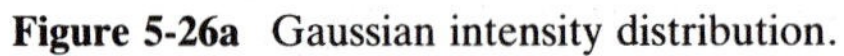

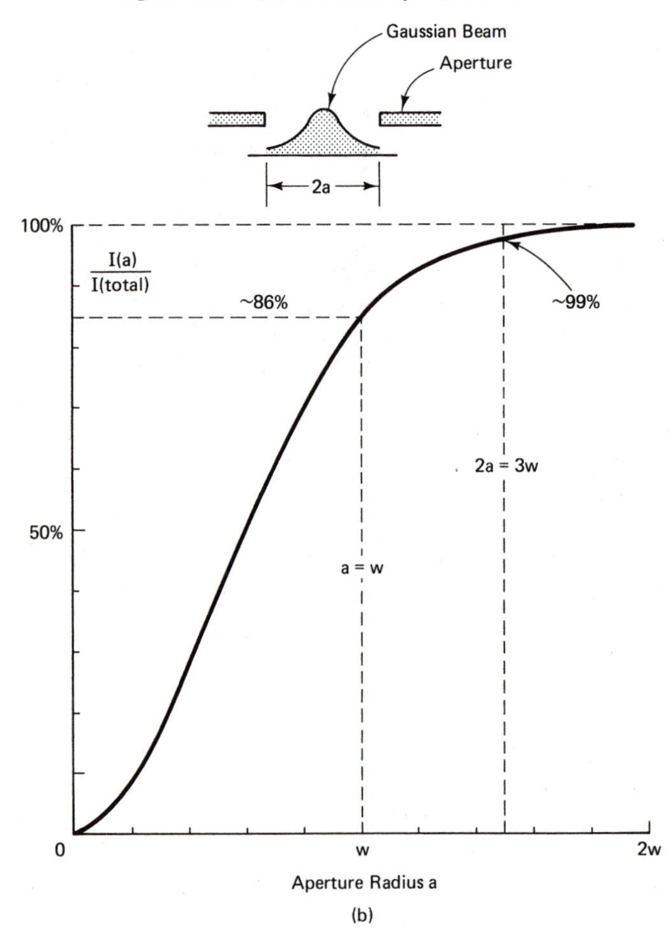

Figure 5-26b Power transmission of a Gaussian beam of spot size w through a circular aperture of radius a.

Another property, which started the practice of specifying beam spread in the far field by its spot size, is that the diffracted Gaussian beam is just another enlarged Gaussian beam when there is little or no lateral truncation of the beam (Figure 5-27). In that case, the diffracted spot radius has been shown to be.[1]

$$W(Z) = W_o^2\left[1 + \left(\frac{\lambda Z}{\pi W_o^2}\right)^2\right] \approx \frac{\lambda Z}{\pi W_o} \tag{5-7}$$

where W_o = minimum spot size at which the wavefront is always a plane; this is also called the waist radius or size.

We now can appreciate the beam spread specification notation of FWe^{-2}, which means full angular width with boundary at the spot radius or the $1/e^2$

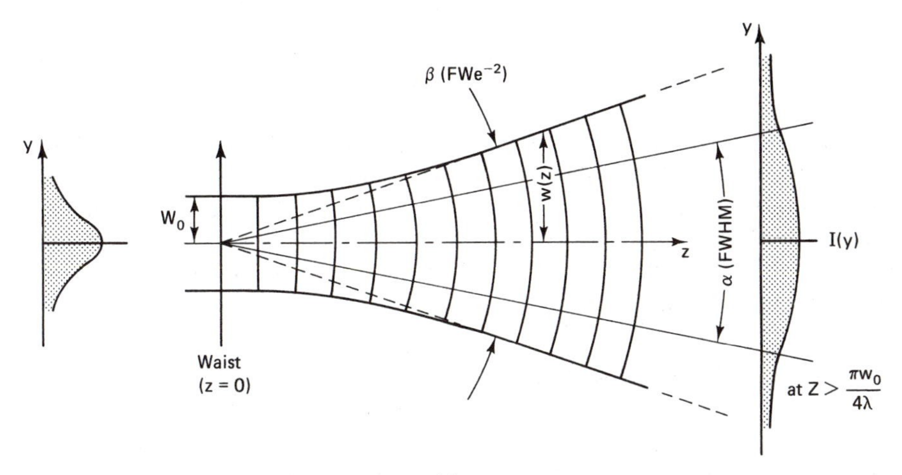

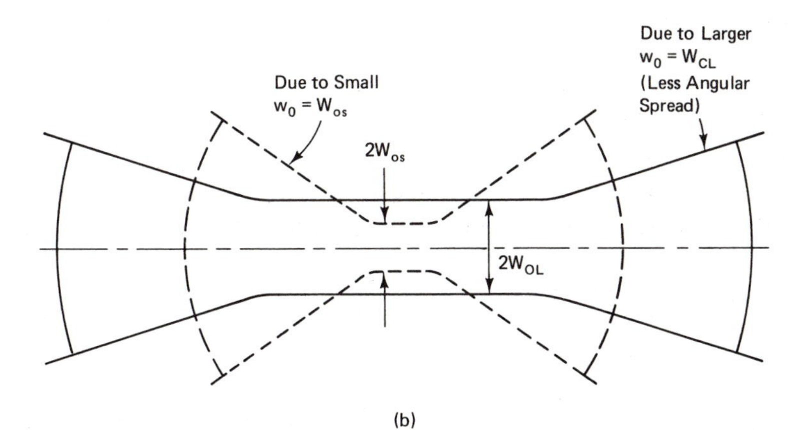

Figure 5-27 Schematic of Gaussian beam propagation along the Z-axis.

intensity point. For example, for untruncated diffraction the far-field beam spread is simply

$$\beta(FWe^{-2}) = \frac{2\lambda}{\pi W_o} \tag{5-8}$$

Notice that the corresponding beam spread (α) at full width half maximum (FWHM) is smaller than β (Figure 5-27) by a factor of 0.5886.

Another aspect worth reviewing is that the wavefront radius (R) of a Gaussian beam could change rather substantially and continuously as it propagates over a distance of only a few inches, contrary to popular notions about laser beams, depending on the spot radius W. In short, a small beam diameter will manifest more diffraction spread in a shorter distance. For an untruncated beam, the following equation relates R to W_o:

$$R = -Z\left[1 + \left(\frac{\pi W_o^2}{\lambda Z}\right)^2\right] \tag{5-9}$$

To get high efficiency in energy coupling from the laser output to the other optics, only the lowest cavity mode designated as TEMoo (Figure 5-28) is suitable, and only the TEMoo mode emits a Gaussian beam. For all practical purposes, the higher mode patterns shown (Figure 5-28) are multiple independent light sources which obviously do not couple efficiently or do not collimate adequately (Sec. 5.2.2).

5.8.2 External Propagation

It is useful to divide the laser beam propagation region into an internal and an external region of the transmitter system in order to take full advantage of the pertinent analytical tools developed. This section begins with external propagation since it relates directly to the general diffraction theory, which we will briefly review.

Diffraction Theory

For most optical antenna, the Fresnel-Frannhofer diffraction integral of the following form will be precise enough to handle any geometry, intensity distribution, and wavefront aberration of engineering interests (Figure 5-29).

$$I(u,v) = \frac{A}{\lambda^2 Z^2}\left|\iint U(x,y)\exp\left[-j\frac{2\pi}{\lambda Z}(xu + yv)\right]dx,dy\right|^2 \tag{5-10}$$

where

$$U(x,y) = E(x,y)\,\frac{e^{-ik}(R_o + \Delta(x,y))}{R_o} \tag{5-11}$$

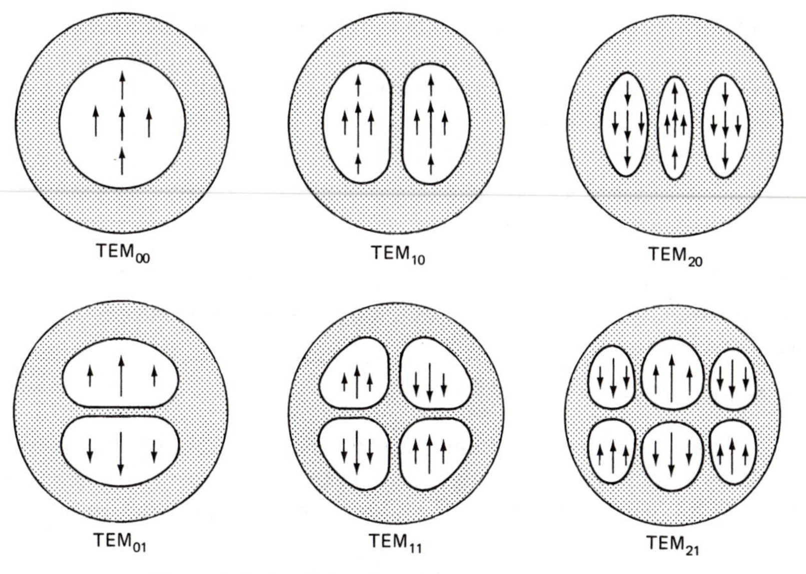

Figure 5-28 Spatial mode pattern of the laser output.

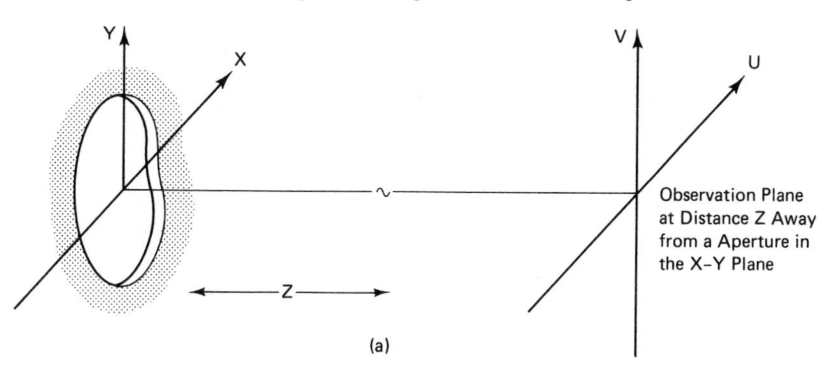

Figure 5-29a Coordinates for diffraction integral.

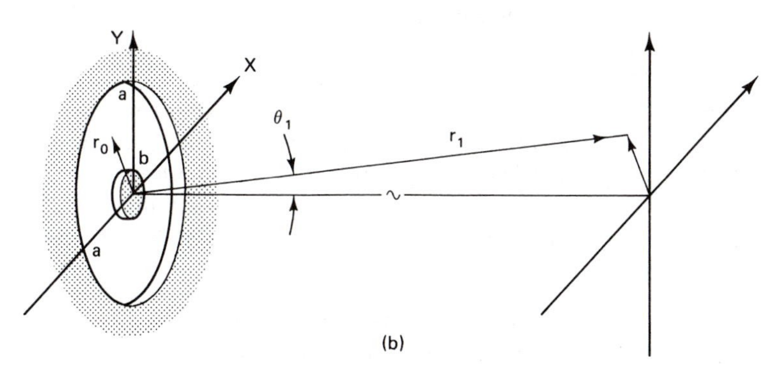

Figure 5-29b Coordinates for diffraction integral with circular symmetry.

$E(x,y)$ = arbitrary amplitude distribution across the aperture
R_o = radius of the wavefront at the exit plane of the telescope
$\Delta(x,y)$ = aberration in form of wavefront deviation from R
λ = wavelength
A = area of the arbitrary aperture
k = $2\pi/\lambda$
x,y,u,u = coordinate axes shown in Figure 5-29a

Although Equation (5-10) can be calculated efficiently by fast Fourier transform algorithm,[6] this two-dimensional integral often is not necessary in this application because of symmetry.

Antenna Gain

Circular symmetry in antenna aperture and wavefront distribution is the most often encountered case in the crosslink transmitter, in which case Equation (5-10) can be turned into a Hankel transform of the following form:

$$I(r_1, \theta_1) = \frac{k^2}{r_1^2} \left| \int_b^a E_0(r_0) \exp\left(j \frac{kr_0^2}{2r_1} \right) J_0(kr_0 \sin \theta_1)\, r_0 dr_0 \right|^2 \tag{5-12}$$

where

$$E_0(r_0) = \left(\frac{2}{\pi W^2} \right)^{1/2} \exp\left(\frac{-r_0^2}{W^2} \right) \exp\left(\frac{jkr_0^2}{2R} \right) \tag{5-13}$$

where

a = aperture radius
$E_0(r_0)$ = laser's Gaussian amplitude function
r_0 = radius in the plane containing the obscuration radius b
k = $2\pi/\lambda$
R = wavefront radius in the aperture
λ = wavelength
r_1 = range to observation point
θ_1 = off-axis angle (Figure 5-29b)
J_0 = Bessel function

The limits of this integral account for truncation by primary mirror radius (a) and obscuration by secondary mirror radius (b), the net results are loss in power transmitted by the telescope plus more beam spreading.

Since similar losses are insignificant in microwave antenna, a subtle modification in the definition of microwave antenna gain is necessary[7] before applying it to the optical antenna. That is:

$$G_T = \frac{\text{Power density at target from antenna}}{\text{Power density at target from isotropic source with the same input power}} \tag{5-14}$$

The subtle change to input power (P_i) does not affect the microwave case where almost always the output equals input power, but it does account much more realistically for the optical antenna gain. When relating Equation (5-14) to Equation (5-12), the optical antenna gain can be more specifically written as

$$G_t(r_1, \theta_1) = \frac{I(r_1, \theta)}{I_0} \tag{5-15}$$

where

$$I_0 = \frac{P_i}{4\pi r_1^2} \tag{5-16}$$

and

$$\int_0^{2\pi} \int_0^{\infty} |E_0(r_0)|^2 \, r_0 dr_0 d\varphi_0 = P_i \tag{5-17}$$

Yet, using algebra, Equation (5-15) can be changed into a convenient form (Equation 5-18) perhaps more familiar to some readers[8]:

$$G_t(r_1, \theta_1) = \frac{4\pi A}{\lambda^2} \, g_t(r_1, \theta_1) \tag{5-18}$$

where $A = \pi a^2$, and $4\pi A/\lambda^2$ is the well-known gain of a circular and uniformly illuminated microwave or optical antenna with no obscuration or aberration. The function g_t is an integral which is almost identical to Equation (5-12) except for a few constants; that is:

$$g_t = 2\alpha^2 \left| \int_{\gamma^2}^{1} e^{i\beta h} e^{-\alpha^2 h} \, J_0(X\sqrt{h}) \, dh \right|^2 \tag{5-19}$$

where:

$h = (r_0/a)^2$
$\alpha = a/w$ = truncation ratio
$\gamma = b/a$ = obscuration ratio
$X = ka \sin \theta_1$ = angular extent
$\beta = ka^2/2 \, (1/r_1 + 1/R)$ = defocus parameter

In view of the definition of gain, it is easy to associate a high gain antenna with narrow beam spread. It also is easier to compare optical with microwave antenna and they both can use this same range equation to calculate the power (P_r) received as follows:

$$P_r = \frac{P_i G_t G_r \lambda^2 T_t T_r}{(4\pi r_1)^2} \tag{5-20}$$

where

P_i = input power
T_t = transmitter transmittance
T_r = receiver transmittance
G_r = receiver antenna gain

$$G_r = \frac{4\pi A}{\lambda^2}(1 - \gamma^2) \tag{5-21}$$

$$\gamma = \text{obscuration ratio} = b/a$$

A few graphs that the author finds useful for optical antenna design are shown as Figures 5-30 through 5-33. They are calculated from Equation (5-19) by varying various parameters of interest to antenna design.

Small Aberration

Even a well-designed transmitter has some small aberration and to estimate its impact on antenna gain, there is fortunately, a short-cut [otherwise Equation (5-10) has to be used]. The definition of Strehl ratio(s) is:

$$S = \frac{I'_{pk}}{I^{\circ}_{pk}} \tag{5-22}$$

where
I°_{pk}

I'_{pk} = intensity at the peak of a far field pattern from an actual beam which usually has some aberration.
I°_{pk} = intensity at the peak of a similar far field pattern from an ideal beam which has no aberration (usually calculated)

As shown by Ref. 8, when the aperture is clear and the amplitude distribution is uniform within it, the Strehl ratio can be related to rms wavefront error (σ), independent of the nature of the aberrations, as follows:

$$S = e^{-(k\sigma)^2} \tag{5-23}$$

where

$$\sigma \leq \lambda/5,\ k = 2\pi/\lambda$$

To determine the impact of violating the strict caveats stated above, Equation (5-11) is used to calculate what is being approximated by Equation (5-23) and is shown on Figure 5-34. Those values at the intersections of vertical axis and curves, which represented small aberration $\leq\lambda/5$, small obscuration ratio <0.3, and small encircled area centered on the peak value, are very close to that calculated by Equation (5-23). Therefore, the simple Equation (5-23) for

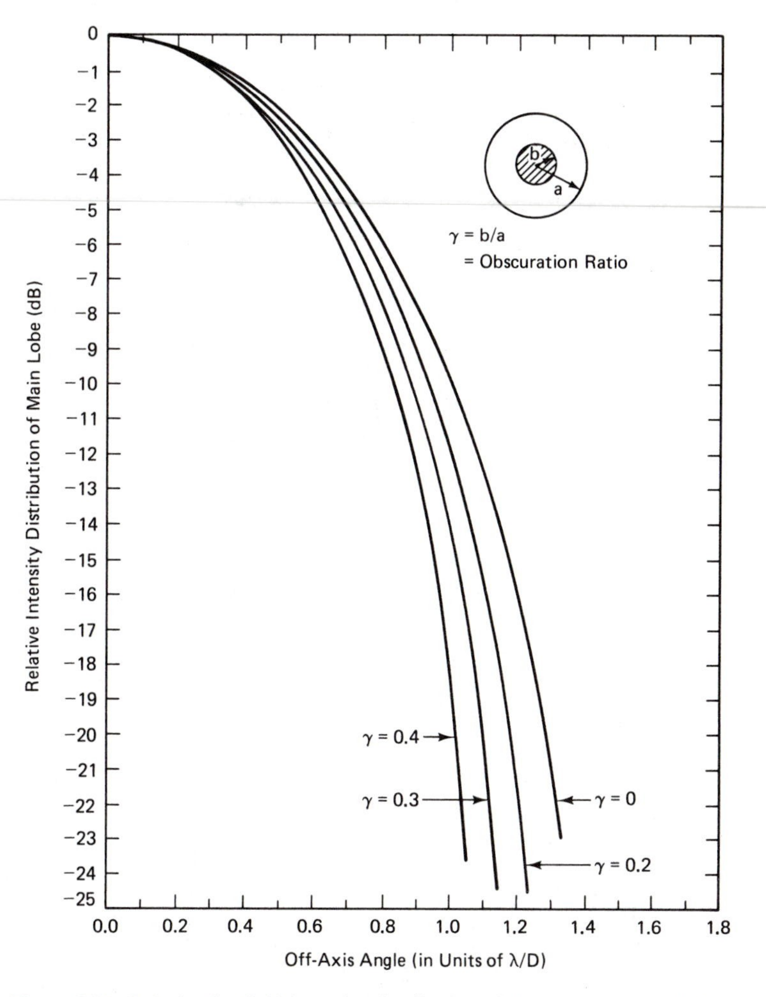

Figure 5-30 Relative far-field intensity distribution of the main antenna lobes (normalized individually).

Strehl ratio is indeed a useful engineering tool even when there is some obscuration (≤ 0.3) and Gaussian amplitude distribution across the aperture.

However, Equation (5-23) often is abused by applying it to a large encircled area centered on the peak. According to Figure 5-34, the valid circle for Equation (5-23) to be accurate, should be no bigger in diameter than 25 percent of the beam spread (FWe^{-2}) of an aberration free wavefront. The approximation in form of Equation (5-23) will produce invalid undervalued ratio if the encircled area is larger than stated above. This constraint along with $\sigma \leq \lambda/5$ often are ignored, thus leading to many erroneous conclusions regarding wavefront quality based on Equation (5-23).

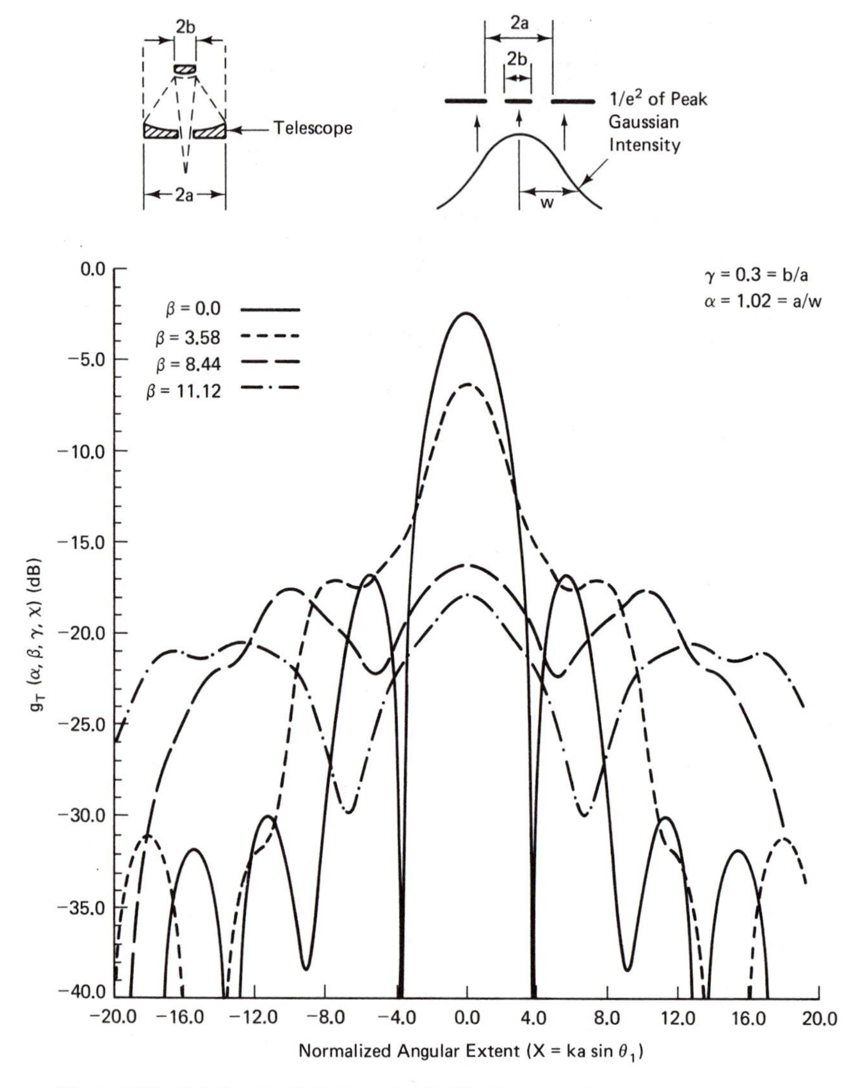

Figure 5-31 Relative far-field intensity distribution spanning many antenna lobes. See Equation (5-18). β is antenna defocus parameter.

5.8.3 Internal Propagation

One of the major functions of the IOS is to propagate the laser beam to the exit plane (the plane containing the secondary mirror) of the telescopes efficiently and with the desirable characteristics for maximum axial gain, such as shown on Figure 5-32. They are summarized as follows:

1. A plane wave in the exit plane with little or no aberration.

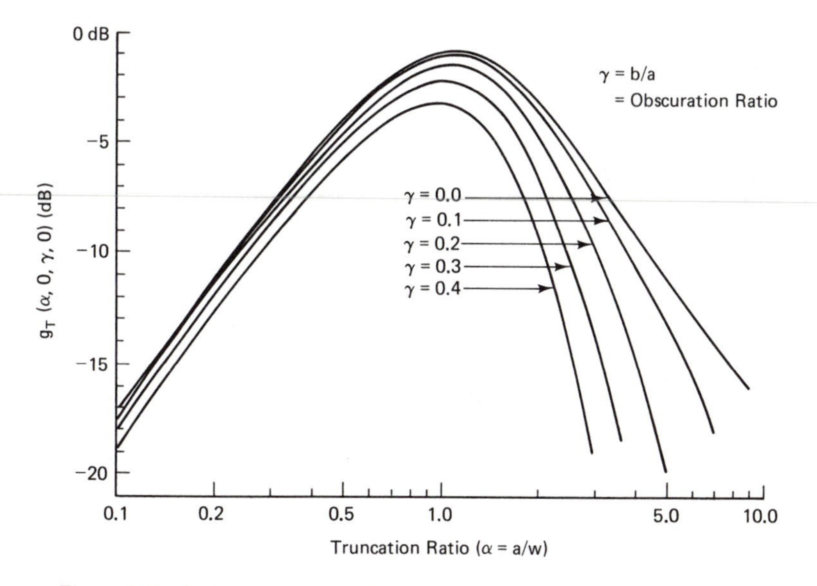

Figure 5-32 Optimizing antenna far-field axial gain. See Equation (5-8).

2. The truncation ratio, a/w, should be about 1.12.
3. The obscuration ratio, b/a, should be as small as possible.

To accomplish all these, geometrical optics codes and methods will be pressed into action first. However, the basic nature of Gaussian beam propagation (Sec. 5.7.1) will often foil the geometrical design by not having the desired degree of collimation or focus. These discrepancies can be ameliorated in two or three iterations involving geometrical and Gaussian beam methods. The later method is a very efficient and accurate means of obtaining waist radius and its position in a well corrected optical system. This useful technique will be discussed in the rest of this chapter.

Gaussian Beam Method

In principle, Equation (5-10) can be applied to each optical element to find out exactly how the Gaussian beam has propagated; however, this is extremely expensive in practice. Kogelnik and Li had derived algebraic equations to describe lens-to-lens Gaussian beam propagation which greatly improved and simplified the Gaussian beam shaping process.[9] Here, the word lens and mirror can be used interchangeably.

Since in wave optics, a wavefront has both finite transverse dimension and wavefront radius, it is not surprising that the lens equations for a Gaussian beam will include them in addition to wavelength and lens power. As such, the important parameters (Figure 5-35) are now waist positions (d_1, d_2) and

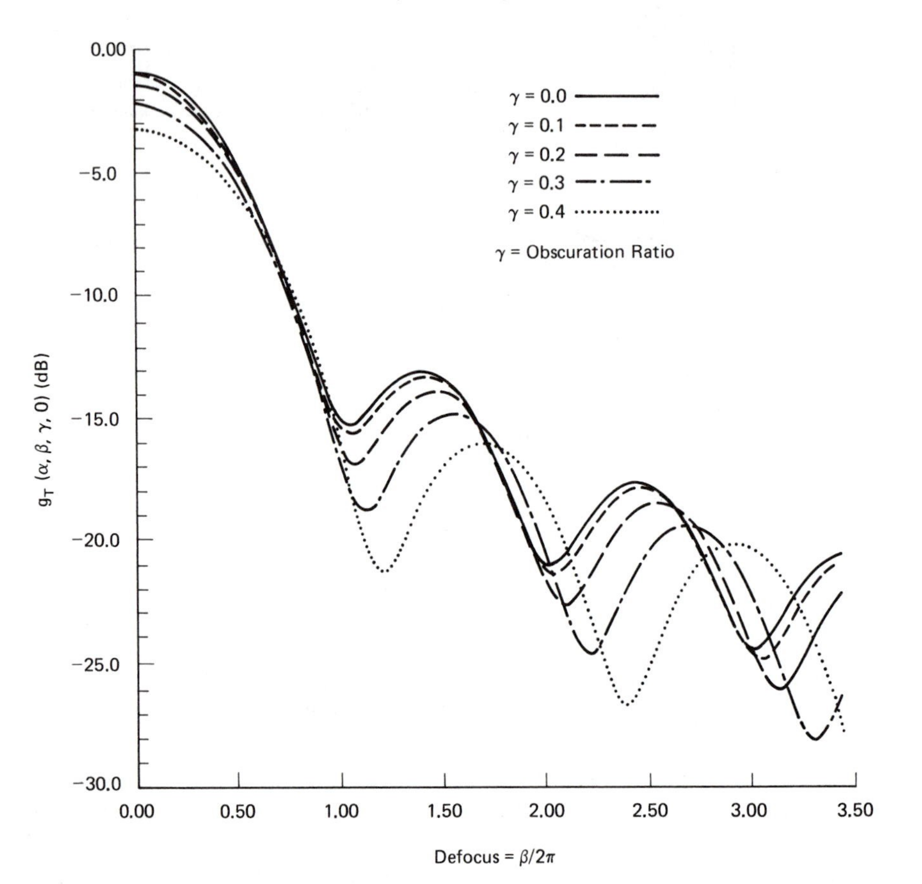

Figure 5-33 Relative far-field axial intensity (axial gain) of a laser antenna as a function of defocus parameter β. See Equation (5-19).

waist radii (W_{01}, W_{02}) before and after each optical element. Specifically, the equations are

$$\frac{d_2}{f} - 1 = \frac{\left(\dfrac{d_1}{f} - 1\right)}{\left(\dfrac{d_1}{f} - 1\right)^2 + \left(\dfrac{b_1}{2f}\right)^2} \tag{5-24}$$

where

d_2 = waist position in image space of optical element
f = focal length of an optical element (+ sign should be used for concave mirror)
d_1 = waist position in the object space of optical element
b_1 = $2\pi W_{01}^2/\lambda$

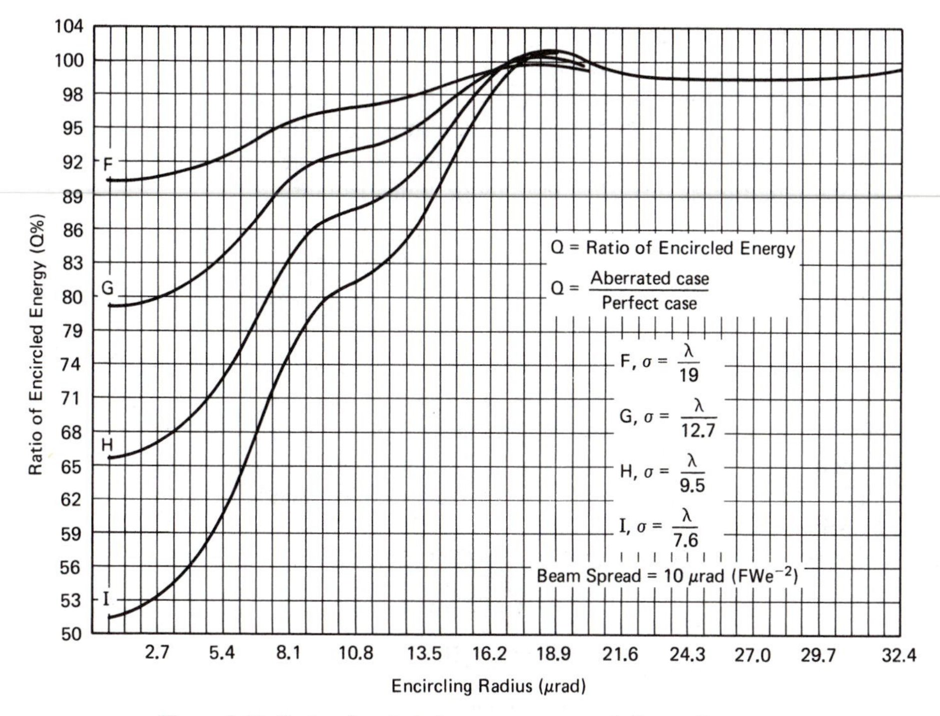

Figure 5-34 Ratio of encircled energy versus encircling radius.

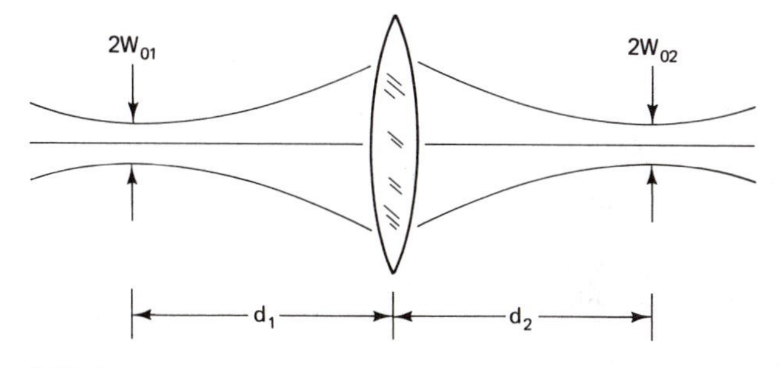

Figure 5-35 Important parameters of a Gaussian beam propagating through a lens or a mirror.

W_{01} = waist radius in the object space of an optical element
λ = wavelength

d_2, d_1, f all have standard sign convention as for lens and mirror.

An important assumption made in Equation (5-24) is that the limiting aperture radius (a) of an optical element is much larger than the spot radius (W) of the incident Gaussian beam at the lens. As a rule, this means that the accuracy of Equation (5-24) will slowly degrade unless the ratio $a/W \geq 1.5$

is maintained at each optical element, especially if the beam is to encounter a number of optical elements.

As the output waist radius (W_{02}) becomes the input waist radius (W_{01}) of the following lens, the transformation of waist radius by an optical element must be known because it influences the waist position. Therefore the following equation is given[9] as the law of transformation:

$$W_{02}^2 = \frac{W_{01}^2}{\left(\frac{d_1}{f} - 1\right)^2 + \left(\frac{b_1}{2f}\right)^2} \tag{5-25}$$

As a numerical example using the above two equations and Equation (5-9), one can perturb the d_1 by a couple of microns in order to find the change in wavefront radius at the output plane of a Cassagrain telescope shown on Figure 5-16. Because the calculation in this case requires seven to eight significant digits, one should use a good calculator or a computer. Consider the following example:

Primary mirror focal length = 379.5225 mm

$$\lambda = 1.06 \times 10^{-3}\text{mm}$$

$$W_{01} = 1.69596 \times 10^{-3}\text{mm}$$

$$\Delta d_1 = 2 \times 10^{-3}\text{mm (separation error)}$$

$$d_1 = 379.5246 \text{ mm} + \Delta d_1$$

$$f = 379.5225 \text{ mm}$$

Equation (5-24) will give $d_2 = 3.924502 \times 10^6$ mm. However, we want to know the exiting wavefront radius (R) at the primary mirror so letting $Z = d_2$ and $W_o = 75.7886$ mm (Equation (5-25)), we get $R = 77.7$ km by using Equation (5-9). This is a big reduction from $R = 2554$ km when $\Delta d_1 = 0$.

Figure 5-16 is obtained by the above method.

5.9 CONCLUDING REMARKS

In this chapter, detailed discussions were given of the constituent components and methods used to synthesize and analyze the optics of a laser satellite crosslink. For the sake of completeness, some basic tutorial material was included. The unique design requirements were addressed and the challenge and limitations facing the design engineer were given.

REFERENCES

1. A. E. Siegman, "An Introduction to Lasers and Masers" (New York: McGraw-Hill Book Company, 1971).
2. E. Hecht and A. Zajac, *Optics,* (Reading, Mass: Addison-Wesley, 1976).
3. N. Feazel, "Shuttle IPS Unit," *Aviation Week*, pp. 155–56, Aug. 13, 1984.
4. R. F. Begley, D. Goux, D. W. Chan, and O. Giat, "Laser Beam Combiner," *Applied Optics*, Vol. 21, No. 17, pp. 3213–20. Sept. 1982.
5. C. McIntyre, W. N. Peters, C. Chi, and H. Wischina, "Optical Components and Technology in the Laser Space Communications Systems," *Applied Optics*, Vol. 58, No. 10, pp. 1491–1503, Oct. 1970.
6. J. W. Goodman, "Introduction to Fourier Optics," (New York: McGraw-Hill Book Company, 1968).
7. B. J. Klein and J. J. Degnan, "Optical Antenna Gain," *Applied Optics*, Vol. 13, pp. 2134–42, Sept. 1974.
8. M. Born and E. Wolf, *Principles of Optics*, Oxford: Pergamon, 1965.
9. H. Kogelnik, T. Li, "Laser Beams and Resonators," *Proc. IEEE*, Vol. 54, pp. 1312–29, Oct. 1966.

6

Laser Beam Pointing Control Acquisition and Tracking Systems

J. M. Lopez and Dr. K. Yong
The Aerospace Corporation, El Segundo, CA

6.1 INTRODUCTION

Acquisition, tracking, and pointing (ATP) of a laser beam in space plays an important role in the space-based, laser communication (lasercom) system. The very narrow beam widths together with the margins required for communication under various environmental conditions impose critical pointing requirements that present some unprecedented challenges to control engineers.

This chapter addresses the various acquisition methods and critical technological issues of tracking and pointing for satellite-based laser communication systems, under their dynamic operating environment and stringent mission requirements. The most stressing design parameters and components for various applications are discussed. The structure of a digital computer simulation program that can be utilized for performance verification also is given.

A schematic diagram of an example of a space lasercom relay network is shown in Figure 6-1. In this concept, spacecraft *A* transmits data via a laser beam to spacecraft *B*, while spacecraft *B* sends signals to the ground station. Identical design of the lasercom subsystem is possible for spacecrafts *A* and *B* thereby reducing the cost of development. Among all of the proposed applications the one common function required of a lasercom subsystem is that it must first acquire the opposite spacecraft. After acquisition it must

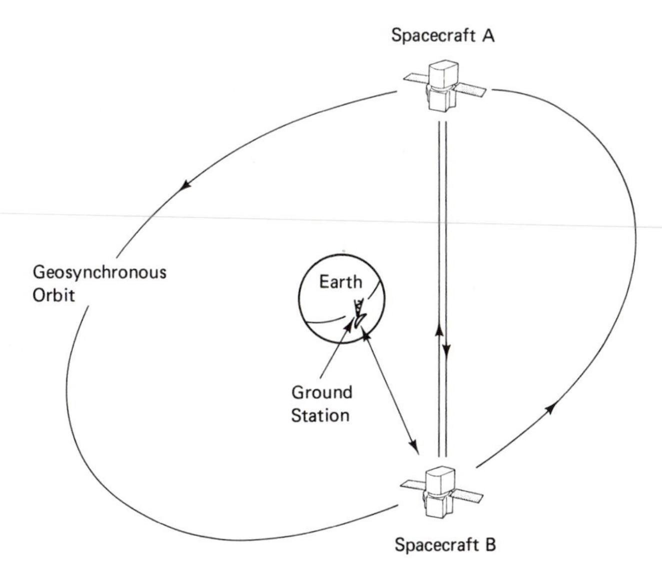

Figure 6-1 Schematic diagram of a lasercom relay network.

maintain closed-loop tracking and pointing with great accuracy for intervals of time that could last as long as several months or years. If, for any reason, this tracking and pointing control link is broken, reacquisition must be conducted as rapidly as possible to avoid further communication interruption. It is obvious that the acquisition, reacquisition, tracking and pointing control technology generally referred to as ATP technology is a critical issue for all space laser applications. A typical ATP performance requirement specification for future communication satellites using a lasercom system is given in Table 6-1.

As indicated in Table 6-1, range separation between two geosynchronous satellites can be about 84,000 km. In this case, a lead-ahead angle compensation, due to the finite speed of light, becomes necessary in presteering the control optics toward the velocity vector of the receiving vehicle. Due to longer lifetime requirements for future spacecrafts, subsystem components must demonstrate a seven- to ten-year 2σ reliability. This constraint also imposes a stringent requirement on the current state-of-the-art control system components and sensors.

6.2 SYSTEM DESCRIPTION

The lasercom subsystem for a spaceborne communications link can be divided into seven major functional blocks as indicated in Figure 6-2. The seven major subsystems are (1) the coarse steering gimbal assembly, (2) fine pointing

TABLE 6-1 PARAMETERS FOR FUTURE COMMUNICATION SATELLITES USING A LASERCOM SYSTEM

Functions	Requirements
Host vehicle control	3 axes active
Spacecraft attitude determination accuracy	180 μrad (2σ)
Range separation	84,000 km
Dynamic rate	18,000 μr/s
Coarse pointing accuracy	10 to 100 μrad (1σ)
Fine pointing accuracy	1 μrad (1σ)
Lasercom LOS (line-of-sight) jitter stability	0.25 μrad (1σ)
Lasercom acquisition time	10 s
Reacquisition time	0.5 s
Estimated system lifetime for 2σ reliability	10 yr.

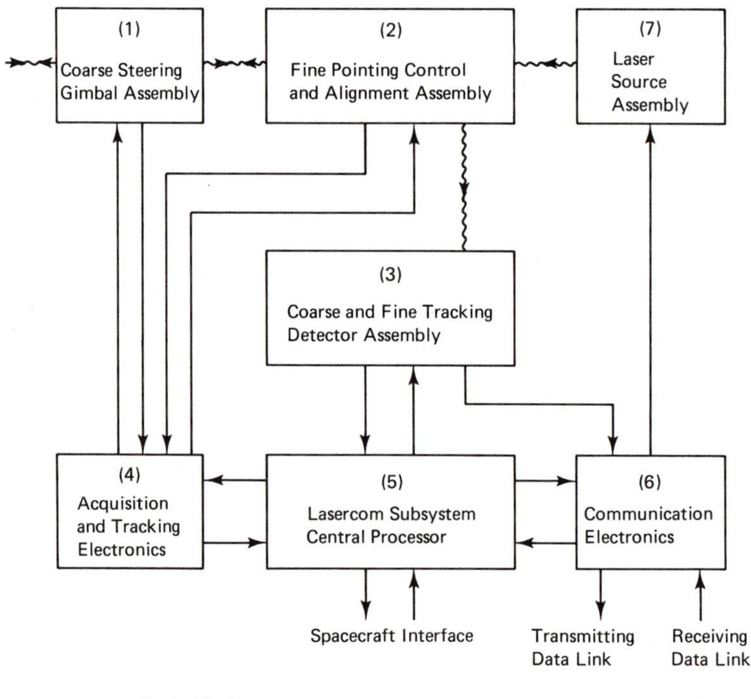

Figure 6-2 Lasercom subsystem functional schematic diagram.

control and alignment assembly, (3) coarse and fine tracking detector assembly, (4) acquisition and tracking electronics assembly, (5) lasercom subsystem central processor, (6) communications electronics assembly and (7) laser source assembly. The related components and technology development of the first five subsystems will be addressed in this chapter. The last two subsystems are discussed in Refs. 1 through 5 and will not be covered.

6.2.1 Coarse Steering Gimbal Assembly

The coarse steering gimbal assembly typically consists of the following components: a two axes gimballed telescope (Figure 6-3) or flat mirror (Figure 6-4), relay optics assembly, gimbal angle pick-off sensing devices, and gimbal servo drive motors. During acquisition, the coarse steering gimbal assembly receives open loop gimbal angle command and possibly rate command signals from the acquisition and tracking electronics to steer the received broad acquisition laser beam into the acquisition detector's field of view (FOV). During fine tracking, this assembly ensures that the received laser beam is within the fine tracking and pointing control system dynamic range. In general, the noise induced by coarse gimbal bearing friction is one of the most important disturbances which affects the lasercom pointing and the attitude determination and control of the host spacecraft. The technology of the coarse steering gimbal system is reasonably mature; however, it must be designed individually for each proposed communication satellite because of different mission requirements and the host space vehicle's attitude control system. Commonality of design in the gimbals, the relay optics, and pick-off sensor

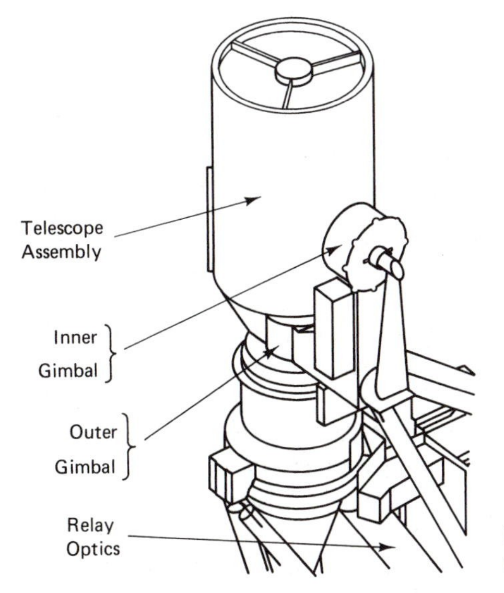

Figure 6-3 Gimballed telescope configuration.

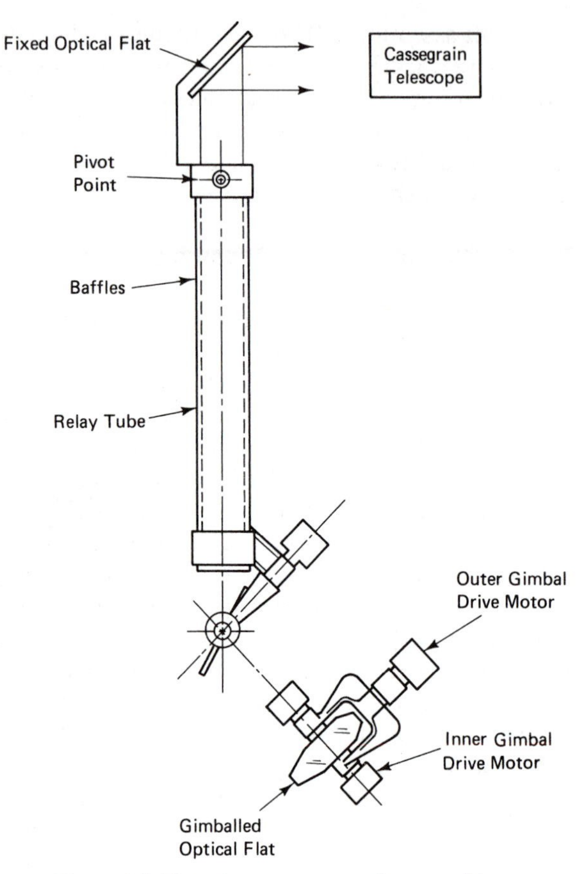

Figure 6-4 Flat mirror coarse-steering assembly.

is quite practical; however, the control servo of the coarse loop could be different for different communication satellites.

6.2.2 Fine-Pointing Control and Alignment Assembly

The fine-pointing control and alignment assembly performs the critical lasercom pointing function in the laser communication system. Depending on mission and range requirements, it may consist of several sets of gimballed optics and their associated torque motors to perform the fine pointing, look-ahead compensation, and boresight alignment functions. After handover from the acquisition mode, fine pointing usually is accomplished in a closed-loop fashion via error signals from the fine-tracking detector. High control-loop bandwidth in the order of 200 to 300 Hz usually is required to remove laser beam jitter and provide fast response. Depending on the frequency and accuracy of ephemeris and target position information, a velocity propagator or a tracking estimator may be required to compute the lead-ahead angle. The lead-ahead angle compensation usually is performed in an open-loop fashion. To ensure precision pointing accuracy, on-board boresight alignment and calibration are necessary. For jitter stability control and to provide attitude reference information, state-of-the-art gyros or accelerometers may be needed. It is possible to develop a common fine pointing and alignment assembly subsystem for various future laser communication satellites. Redundancy may be required to provide a ten-year 2σ reliability for this type of assembly.

6.2.3 Coarse-and Fine-Tracking Detector Assembly

The tracking detector assembly generally consists of an acquisition coarse detector, a fine-tracking detector, and a data processor and its associated electronics. The acquisition detector has a wider detector field of view (FOV) and provides error signals for the coarse-steering gimbal assembly while the fine-tracking detector has a smaller FOV and provides tracking error signals for the fine-pointing and alignment assembly. Quadrant detectors can be used for the purpose with a wide range of data-processing rates depending on the acquisition and tracking methodology used. This will be discussed in detail in the next section. The signal-to-noise ratios of these detectors and their data-processing rates are a dominant factor in achieving desired fine-pointing accuracy and thus must be analyzed carefully in the development phase. The technology of the detector assembly development has progressed very rapidly in recent years and no specific development difficulty is anticipated. Commonality is possible for the detector assembly for various future space applications.

6.2.4 Acquisition and Tracking Electronic Assembly

The acquisition and tracking electronic assembly is the brain required to perform the lasercom acquisition, reacquisition, and track and point functions. Information provided by the central processor and the coarse- and fine-tracking detectors command the coarse steering gimbal assembly and the fine-pointing and alignment assembly to perform open-loop acquisition and closed-loop tracking, alignment, and calibration of the entire subsystem. Using spacecraft attitude information and ephemeris data, it also computes the lead-ahead angle to conduct presteering compensation. Since its functions are highly dependent on mission and spacecraft interface requirements, commonality is difficult to achieve. However, the development of a common processor that is able to handle most of the software functions is feasible.

6.2.5 Lasercom System Central Processor

A central processor is required to serve as the interface between the host vehicle and the lasercom system for control and communication functions. Processing of attitude control and determination information and spacecraft ephemeris data are the primary ATP functions of the central processor. Command and communication telemetry also reside in the central processor. Although the residing software could be different from among the various missions because of their different requirements, commonality is possible in developing a processor that can be used for all future missions. Technology development in processor design is advanced and no significant problems are anticipated.

6.3 ACQUISITION METHODOLOGY

6.3.1 Introduction

It is assumed that normal acquisition between the lasercom system elements are self-initiated by the on-board computers of each spacecraft from a previously defined clock and position reference. Furthermore, it is assumed that the acquisition process is not assisted by other conventional communication links (radio, microwave, etc.).

The uncertainty associated with the knowledge of the location (position in orbit) of each spacecraft and the attitude of the lasercom transceiver LOS (line-of-sight) relative to that location form an acquisition error volume for each system. The latter uncertainty is the dominant contributor to the total error volume and its magnitude depends on the spacecraft stabilization system utilized (three-axis, dual-spinner, etc.), quality of inertial measurement sensors (gyros, star sensors, etc.), attitude control system, attitude estimations software, and the disturbance environment.

The usual acquisition technique proposed for laser space communication networks begins with one terminal, say terminal A, (residing on spacecraft A), turning on a beacon for the second terminal, terminal B (residing on spacecraft B), to detect and lock onto. Thus, terminal B acquires the beacon from terminal A. The requirements on the beacon pointing are only that it illuminate terminal B. This may be done with a beacon with a relatively wide divergence angle that fills the entire angular error volume associated with the location of terminal B relative to terminal A, or it may be done with a smaller beam divergence angle that scans the error volume. Note that the communication laser itself can also serve as the beacon, although some broadening (defocus) may be required. The complexity of a scanning technique is avoided if the error volume is small enough so that the beacon can provide an adequate power density over the entire volume, thus providing a good signal for terminal B to detect. However, this condition may be difficult to achieve in practice.

6.3.2 Scanning Techniques

The more general approach will be considered where the beacon cannot fill the entire error volume and terminal B detector FOV is limited, thus both terminals are scanning within their respective error volumes. It is desirable to perform acquisition in a specified minimum time interval; therefore, parameters that significantly affect acquisition time will be identified and discussed. Acquisition will be considered only as that period required for terminal B to recognize and lock-on to the beacon from terminal A. The time required for terminal A to receive and lock-on to a return beacon from terminal B is considerably less than the former. The acquisition scenario also could have both terminals "signaling" and detecting, simultaneously. Figure 6-5 illustrates the geometry of the acquisition problem where

2α = angular cone (error envelope) containing vehicle B with respect to vehicle A

$2\alpha_W$ = beacon beamwidth of terminal A

2β = angular cone (error envelope) containing vehicle A with respect to vehicle B

$2\beta_W$ = detector FOV of terminal B

R = distance between the two vehicles

Square Scanning Technique

It will be assumed that system B is rapidly searching for the beacon emitted by vehicle A. System A is slowly searching with its beacon for vehicle B. It is unknown where vehicles A and B are within their error volumes

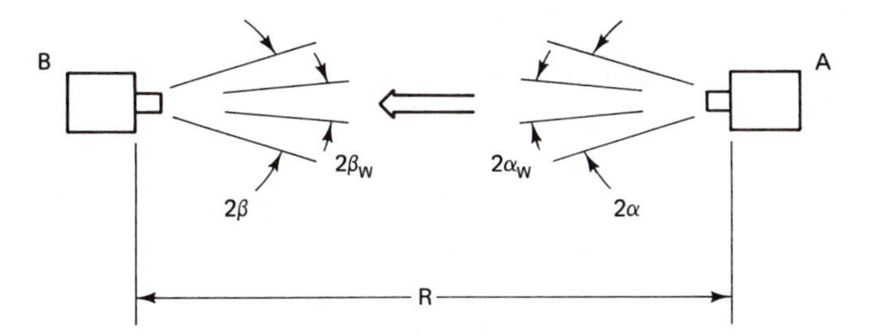

Figure 6-5 Lasercom acquisition geometry.

established by the containment cones (2β, 2α). Further assumptions made are:

1. For both systems, the receiver LOS and transmitter LOS have been boresighted for the predicted lead-ahead angle profile and the detector FOV and beacon beamwidth are larger than the uncertainties associated with this angle.
2. Minimum allowable detector dwell time (t_d) for the fast scanner of system *B* is equal to the inverse of the terminal *A* laser beacon signal repetition rate.
3. Square scanning patterns of areas $(2R\beta)^2$ and $(2R\alpha)^2$ will be used for acquisition as illustrated in Figure 6-6. For pattern I, the fast scanner of system *B* has an angular travel capable of traversing its error volume and sweeps through its error volume once for every time slow scanner A moves one spot (beam diameter) in its own error volume. In the case of pattern II, the fast scanner and fine detector of system *B* has either a limited angular field of travel (FOT) or FOV respectively, and must rely on an additional slow scanner to cover the error volume. In either case, many repeated error volume scans may be required of system *B* before lock-on can be achieved with system *A*. If the error volumes are partitioned into b^2 and a^2 spots respectively ($b = R\beta_W$, $a = R\alpha_W$), for terminals *B* and *A*, then the *ideal maximum* number of spots to be scanned before lock-on is the product of a^2b^2. However, this does not account for effects such as false alarms or the dynamics of the scanner system spacecraft.
4. The telescope internal fast steering lens system is used for fast scanning and an exterior gimballed mirror or the telescope itself is used for the slow scanner. Using the fine-pointing component of the tracking system to do the fast scanning required for acquisition is advantageous because of its availability and the minimal modification required to perform this function.

From the geometry previously illustrated, the following relationships are obtained

$$\dot{\theta}_{T_B} = \frac{2\beta_W}{t_d} \quad \text{(angular velocity of fast scanner of system } B\text{)} \tag{6-1}$$

$$b = \frac{\beta}{\beta_W} \quad \text{(number of scan spots per scan line of system } B\text{)} \tag{6-2}$$

$$T_{S_B} = \left[\frac{\lambda_{\text{FOV}}^2}{\beta_W}\right] \quad \text{(total number of spots within area scanned by fast scanner of system } B\text{)} \tag{6-3}$$

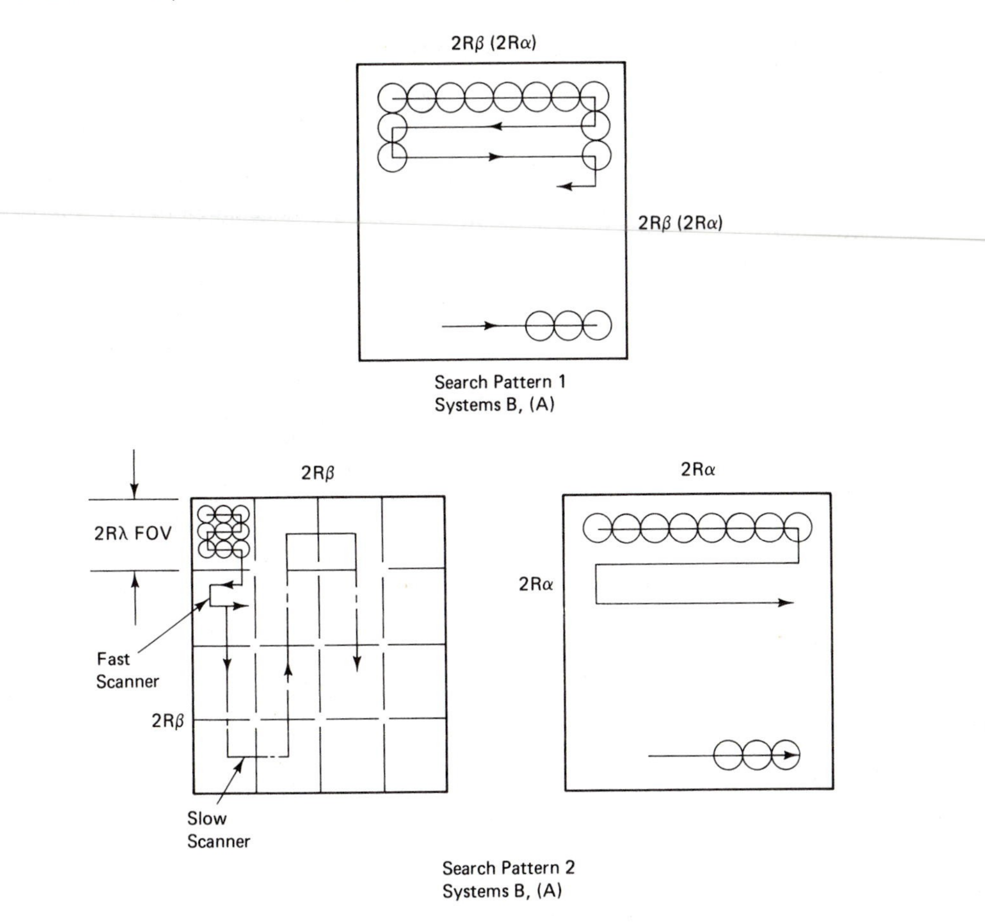

Figure 6-6 Square scanning patterns.

where λ_{FOV} = FOT/2 is half the angular travel of the fast scanner (if $\lambda_{FOV} = \beta$, then slow scanner is *not* required).

$$\dot{\theta}_{S_B} = \frac{2\beta_W^2}{\lambda_{FOV} t_d} \quad \text{(angular velocity of slow scanner of system B)} \tag{6-4}$$

$$a = \frac{\alpha}{\alpha_W} \quad \text{(number of scan spots per scan line of system A)} \tag{6-5}$$

$$t_{\alpha_W} = b^2 t_d \quad \text{(time to move over one spot diameter of system A)} \tag{6-6}$$

$$\dot{\theta}_{S_A} = \frac{2\alpha_W}{b^2 t_d} \quad \text{(angular velocity of slow scanner of system A)} \tag{6-7}$$

The previous relationships are dependent upon the dwell time (t_d), the partitioning of the error volume (b^2, a^2), the FOV of the detector, or FOT of the fast scanner. Determination of values for these parameters is based on communication, detector, and optical requirements.

The total time required to perform acquisition for a square scanning pattern, assuming the maximum number of samples are taken (a^2b^2), can be expressed as

$$t_{\text{acq}} = b^2a^2t_d + k_{fa}t_d + k_tk_{F_{BW}}t_d + k_sk_{S_{BW}}t_d + k_ak_{A_{BW}}t_d \qquad (6\text{-}8)$$

where

k_{fa} = number of "expected" false alarms during acquisition period

k_t = $2ba^2/k + 4k_{fa}$ number of times fast scan loop is accelerated (decelerated) during the scanning procedure where $k = 2_{\text{FOV}}/\beta$ is the ratio of system B fast scanner FOV (or FOT) to its error envelope

k_s = $2a^2/k$ = number of times system B slow scanner is moved (accelerated and decelerated) during the scanning procedure

k_a = $2a$ = number of times system A slow scanner is accelerated (decelerated) during the scanning procedure

ω_{ij} = equivalent natural frequency of the respective scan loops (i = system A or B, j = fast or slow scanner loop)

$k_{F_{BW}}t_d$ = response time of system B of fast scanner loop; approximately equal to $4/\xi\omega_{\text{BF}}$ where ξ is the damping ratio of scan drive loops

$k_{S_{BW}}t_d$ = response time of system B slow scanner loop; approximately equal to $4/\xi\omega_{BS}$

$k_{A_{BW}}t_d$ = response time of system A slow scanner loop; approximately equal to $4/\xi\omega_{AS}$

The first term in the expression is the theoretical maximum acquisition time and can include uncertainties or dynamics of the system. If the detector FOV and fast scanner FOT encompass the error volume, the fourth term can be excluded.

Spiral Scanning Technique

Alternate search patterns which search out the areas of greater contact probabilities within the error volume, such as cruciform or spiral patterns, can also be utilized. A spiral search pattern may actually be better in terms of minimizing acquisition times although a more complicated scan drive signal would be required. Assuming the detector FOV and fast scanner angular travel of system B can encompass its error volume, the ideal scan rate drive signal profile for the two search concepts being considered would be of the form as shown in Figure 6-7. The angular position signal waveform for the square search concept is a periodic triangular wave (X-channel) and alternating ascending and descending staircase wave (Y-channel); and for the spiral search concept the waveform would be alternating damped and diverging sinusoidal oscillations (with a 90° phase shift between the X and Y channels).

The total time required to perform acquisition for a spiral scanning concept, assuming the maximum number of samples are taken ($a_s^2 b_s^2$), can be expressed as

$$t_{\text{acq}} = b_s^2 a_s^2 t_d + k_{fa} t_d + k_{t_s} k_{F_{BW}} t_d + k_a k_{A_{BW}} t_d \tag{6-9}$$

where

$b_s^2 = \pi/4\ b^2$ (number of spots in error volume scanned by systems B
$a_s^2 = \pi/4\ a^2$ and A, respectively)
$k_{ts} = 4k_{fa}$ (number of times fast scanner loop must respond to interrupts)

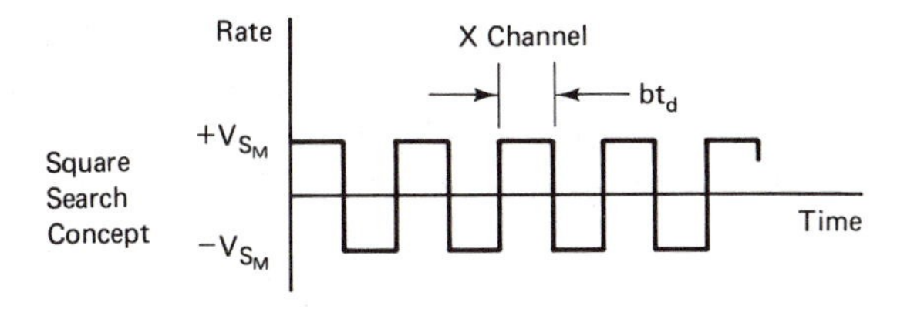

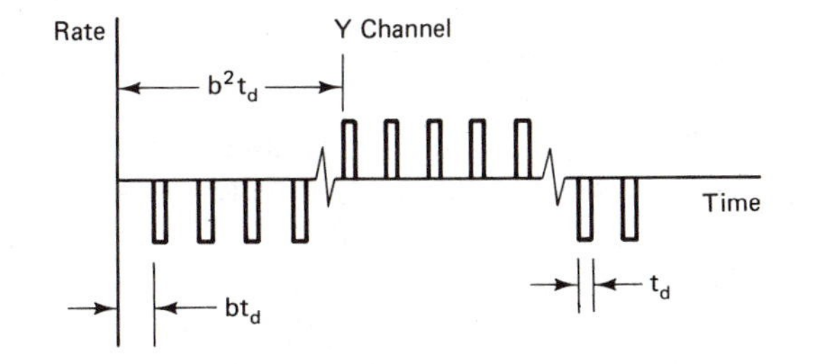

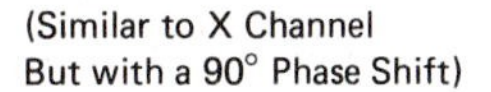

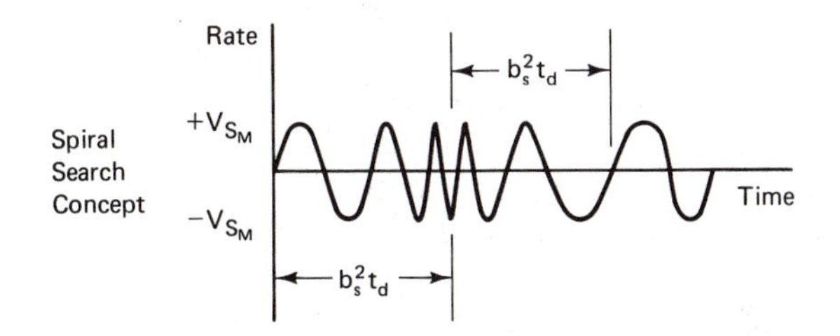

Figure 6-7 Ideal scan rate drive signal profiles.

and the other terms are as previously defined for the square search pattern. The angular scan rates for this concept would be

$$\dot{\theta}_{S_B}(t) = \frac{2\beta}{r(t)_B t_d} \qquad \text{(system } B\text{)}$$

$$\dot{\theta}_{S_A}(t) = \frac{2\alpha}{r(t)_A b^2 t_d} \qquad \text{(system } A\text{)}$$

where $r(t)_i$ is the radial distance from the center of the spiral pattern to the spot being surveyed.

A comparison of the expressions for acquisition times indicates one main difference is in the term $k_t k_{F_{BW}} t_d$. This term is the time delay induced by the response of the fast scanner due to an abrupt signal input. In the square search concept, the signal input is considered abrupt for a false alarm and also when the scan field is changed to the next row or column. However, for the spiral search concept, only the former effect is considered since a relatively smooth transition occurs when the scan field is changing (this is questionable when $r(t)$ approaches its limit). Notice that the difference in search area (square versus circle) or number of spots to be searched can be eliminated by modifying the rate signal for the square search concept so that the corners are not searched. This would reduce the time contribution due to that term by 22 percent.

6.3.3 Basic Design Considerations

Some typical conditions will now be examined to determine their effect on acquisition.

The dwell time should not be less than the inverse of the signal repetition rate, otherwise in the scanning procedure terminal B detector FOV, illuminated by the beacon, may be skipped over. For lasers with pulse-gated modulation signal, repetition rates on the order of 10^4 to 10^5 signals per second can be expected. A dwell time, t_d, of 10^{-4} to 10^{-5} s will be considered. Also, the responsivity of the detector of System B, with its fast scanner looking for the beacon, must be faster than the beacon signal repetition rate.

Partitioning the error volumes into b^2 and a^2 spots should be determined by the following parameters: the amount of beam spoiling (beam widening of the beacon) and receiver FOV allowed (consistent with signal-to-noise ratio), laser power requirements, search overlap required for uncompensated spacecraft drift rates and also by the size of the error volumes to be scanned. The size of the error volume is generally dominated by the attitude uncertainty of the vehicle, and to a lesser extent by the uncertainty of the vehicle location, both of which must be referenced to the telescope or scanner.

The number of false alarms obtained during the acquisition phase depends upon the signal-to-noise ratio and on the length of the acquisition period. However, an estimate of 100 false alarms or less during the period is considered reasonable. A false alarm of one per thousand spots sampled is a realistic design goal. When an alarm occurs (system *B* beacon-detector threshold is exceeded) the scanner discontinues its search pattern and moves back to the spot where the alarm occurred. If during an additional dwell time at that spot another alarm occurs, then the scanner loop is closed and lock-on has begun. If no additional alarm is received, then the previous search pattern is continued and the original alarm is assumed to have been false.

The dynamics of the system are primarily reflected by the slew rates and response times required of each scanner. Because the response time of the fast scanner must be small, the bandwidth of this device must be large. It is for this reason that the fast scanner of the fine pointing system, driven open loop, is considered for scanning. Bandwidths on the order of 200 to 500 Hz have been achieved for these systems with values up to 1,200 Hz predicted. The apparent limitation of these systems is that their angular travel may not encompass the error volumes to be scanned. The field of travel can be increased by combining a slow scanner with the fast system or by using multiple scanners in series.

6.3.4 Typical Initial Acquisition Times

The data of Table 6-2 presents acquisition times for varying conditions previously discussed. Cases 1 through 3 and 7 through 10 reflect the condition where system *B* has a fast scanner FOT and detector FOV which encompasses its error volume (search pattern I) and cases 4 through 6 where it does not (search pattern II). The effects of varying the dwell times and of varying the partitioning of the error volume (which is an indication of beam spoiling) are included in the 10 cases for both square and spiral search patterns. Error volumes of $\alpha = \pm 1.5°$, $\beta = \pm 0.3°$; $\alpha = \pm 0.5°$, $\beta = \pm 0.2°$ and $\alpha = \pm 0.3°$, $\beta = \pm 0.06°$; all 3σ values were considered and should be representative of low earth orbiting systems. General conclusions to be drawn from the data are:

1. Initial acquisition times on the order of 10 s appear to be feasible.
2. FOT and FOV of the fast scanner/detector system to encompass its error volume is highly desirable if not almost mandatory (case 6 shows that for FOT and FOV equal to 1/4 of the error volume ($k = 1/2$) and other factors generally optimized, acquisition time of 10 s can be achieved).
3. Some combination of beam spoiling and error volume reduction is desirable to reduce the amount of partitioning (reduce product a^2b^2 required).
4. The spiral search method tends to reduce acquisition time below the square search method but at the expense of a more complicated drive signal.

TABLE 6-2 ACQUISITION TIMES FOR VARYING DESIGN CONDITIONS

Case No. & Type	$k = \frac{\lambda_{FOV}}{\beta}$	$k_{FBW}t_d$ ~sec	$k_{SBW}t_d$ ~sec	$k_{ABW}t_d$ ~sec	a^2 (SPOTS)	b^2 (SPOTS)	k_t	k_s	k_a	$\dot{\theta}^{(1)}_{TB(2)}$ rad/sec	$\dot{\theta}_{SB}$ rad/sec	$\dot{\theta}^{(1)}_{SA(2)}$ rad/sec	$2\alpha^{(1)}_{W(2)}$ deg	$2\beta^{(1)}_{W(2)}$ deg	$t_{acq.}$ ~sec
1(SQ)	1	7.6×10^{-4}	—	7.6×10^{-4}	100	1000	6800	—	20	3.3	—	0.052	0.3	0.0188	15.19
										0.66	—	0.0104	0.06	0.0038	
2(SQ)	1	1.52×10^{-3}	—	1.52×10^{-3}	100	1000	6800	—	20	3.3	—	0.052	0.3	0.0188	20.35
										0.66	—	0.0104	0.06	0.0038	
3(SQ)	1	1.52×10^{-3}	—	1.52×10^{-3}	81	625	4450	—	18	4.2	—	0.093	0.334	0.024	11.84
										0.84	—	0.018	0.066	0.005	
4(SP)	1/10	1.52×10^{-3}	10^{-2}	10^{-2}	100	1000	64400	2000	20	3.3	1.1	0.052	0.3	0.0188	123.0
										0.66	0.22	0.0104	0.06	0.0038	
5(SP)	1/5	1.52×10^{-3}	10^{-2}	10^{-2}	100	1000	32400	1000	20	3.3	0.525	0.052	0.3	0.0188	69.5
										0.066	0.105	0.0104	0.06	0.0038	
6(SQ)	1/2	7.6×10^{-4}	5×10^{-3}	10^{-2}	81	625	8500	324	18	4.2	0.22	0.093	0.334	0.024	12.33
										0.84	0.044	0.018	0.066	0.005	
*7(SQ)	1	7.6×10^{-4}	—	7.6×10^{-4}	100	1000	6800	—	20	3.3	—	0.52	0.3	0.0188	6.18
										6.6	—	0.104	0.06	0.0038	
*8(SQ)	1	7.6×10^{-4}	—	7.6×10^{-4}	16	900	1360	—	8	23.2	—	0.485	0.25	0.013	1.18
*9(SP)	1	7.6×10^{-4}	—	7.6×10^{-4}	16	900	432	—	8	23.2	—	0.485	0.25	0.01	0.48
10(SP)	1	7.6×10^{-4}	—	7.6×10^{-4}	16	900	432	—	8	23.2	—	0.049	0.25	0.013	1.78

$t_d = 10^{-4}$ sec (10^{-5} sec for cases marked*)

$k_{fa} = 100$

SQ = Square scan

SP = Spiral scan

Slew rates are based on error volumes of: (1) $2\alpha = 3.0°$, $2\beta = 0.6°$; and (2) $2\alpha = 0.6°$, $2\beta = 0.12°$ for cases 1–7.
$2\alpha = 1.0°$, $2\beta = 0.4°$ for cases 8–10.

5. If acquisition times less than 10 s are desired, using a scanning approach, then dwell times less than 10^{-4} s would probably be required. This may be the limiting factor on using the fine pointing system because of the high slew rates required for very small dwell times.
6. Slew rates required for fast scanning and slow scanning are compatible with a fast scanner system and an external gimballed telescope or sweep mirror, respectively.

6.3.5 Establishing the Communication Link

When terminal B has acquired the acquisition beacon from terminal A, it must return a beacon light for terminal A to acquire, thereby completing lock-on. It is assumed the lasers used as beacons for both vehicles have been slightly spoiled to facilitate acquisition and lock-on. Before a communication link can be established, vehicle tracking or pointing errors must be reduced to less than the communication laser beamwidth (in one case approximately 2 μrad). This condition is attained by the closed loop fine pointing tracking system of each vehicle. The time required to accomplish communication link should be less than 0.5 s. The communication laser beams, if used as beacons, can then be focussed and data transmission started.

6.4 TRACKING AND POINTING CONTROL SYSTEM

6.4.1 Introduction

In lasercom applications, one of the terminals transmits a high data rate signal while tracking a low data rate beacon signal and the other terminal tracks the high data rate signal while transmitting the low data rate beacon signal.

The most stringent tracking requirement occurs for the terminal that must point the high data-rate beam, on the order of 1 to 2 μrad to an accuracy of a fraction of its beamwidth.

The fast scanner and servodrive unit used for fine tracking must have the capability of providing boresight alignment of the laser device with the acquisition and tracking detector and associated optics. These critical components should be mounted on a rigid optical bench and possibly isolated from the remaining spacecraft structure if spacecraft-induced disturbances are severe. Periodic boresight calibration of the telescope optics to the tracking detector can be done using star sightings. In addition, since the relative velocity of the two satellites is not zero, a lead ahead angle exists between the receiver LOS and transmitter LOS. This lead-ahead angle varies and can be as large as 80 μrad for some space applications depending on the relative dynamic environment. This is considerably larger than the transmit beamwidth and must be accurately determined.

The target lead-ahead angle and its rate can be estimated by using a tracking estimator (possibly a Kalman filter) that uses measurements from the tracker detector, a laser ranger or ephemeris data and IMU (inertial measurement unit) to obtain the desired accuracy.[9] The tracker detector and IMU are also used to provide error signals to suppress beam jitter induced in the optical path from component-induced disturbances or structural vibrations. Sampling and processing requirements of the tracking detector or self-induced noise generally limit the bandwidth of these sensors. Thus, an IMU capable of providing wide bandwidth sensing up to 100 Hz or greater is required for beam jitter control.

In most lasercom applications, a two servo system would be required where a fine-pointing system (inner loop) is used to attenuate base mount disturbances (frequencies >1 Hz), but because of its limited FOT and fine detector FOV, a coarse-pointing system (outer loop) is required to keep the fine-pointing system operating about a variable null caused by changes in orbit position (frequency <1 Hz).

Figure 6-8 presents a simplified block diagram representative of single axis control of the fine-pointing (inner) tracker loop utilizing a piezoelectric bender actuator to drive the steering element. The fine guidance detector represents, for example, a quadrant detector tube and associated electronics to generate a command voltage to drive the actuator/fast scanner system when actual optical pointing drifts away from the desired null point (θ_{beam}). Some compensation may be required to achieve loop stability and desired response over the operating bandwidth. The fast scanner actuator system can be thought of as a servo-torquer loop with rate and position feedback.

Actual optical pointing is composed of the fine pointing fast scanner loop output (demagnified by K_θ) plus an input from the outer loop telescope drive system (or external view mirror) which basically tries to keep the fine pointing system operating about its null point. The two systems act in unison to track for low frequency outer loop (telescope or view mirror) and high frequency inner loop fast scanner signal inputs. This simple linear representation of the fine pointing tracker system is then used to gain some insight into fine pointing system design when the effect of disturbances induced to the telescope by base motion is considered. A more complete analysis will have to include the effects of nonlinearities such as limiters on the piezoelectric bender drive voltage (Ve) and possibly on the fast scanner position (θ_γ) to avoid depoling, creep, and possible mirror distortion. Also of significance will be the effect of hysteresis in determining and controlling fast scanner position.

6.4.2 Basic Control Design Considerations

Control laser terminals required for most space links will be relatively small. Six- to ten-inch optics are proposed for laser communication subsystems; however, since the telescope system is designed to be nearly diffraction-

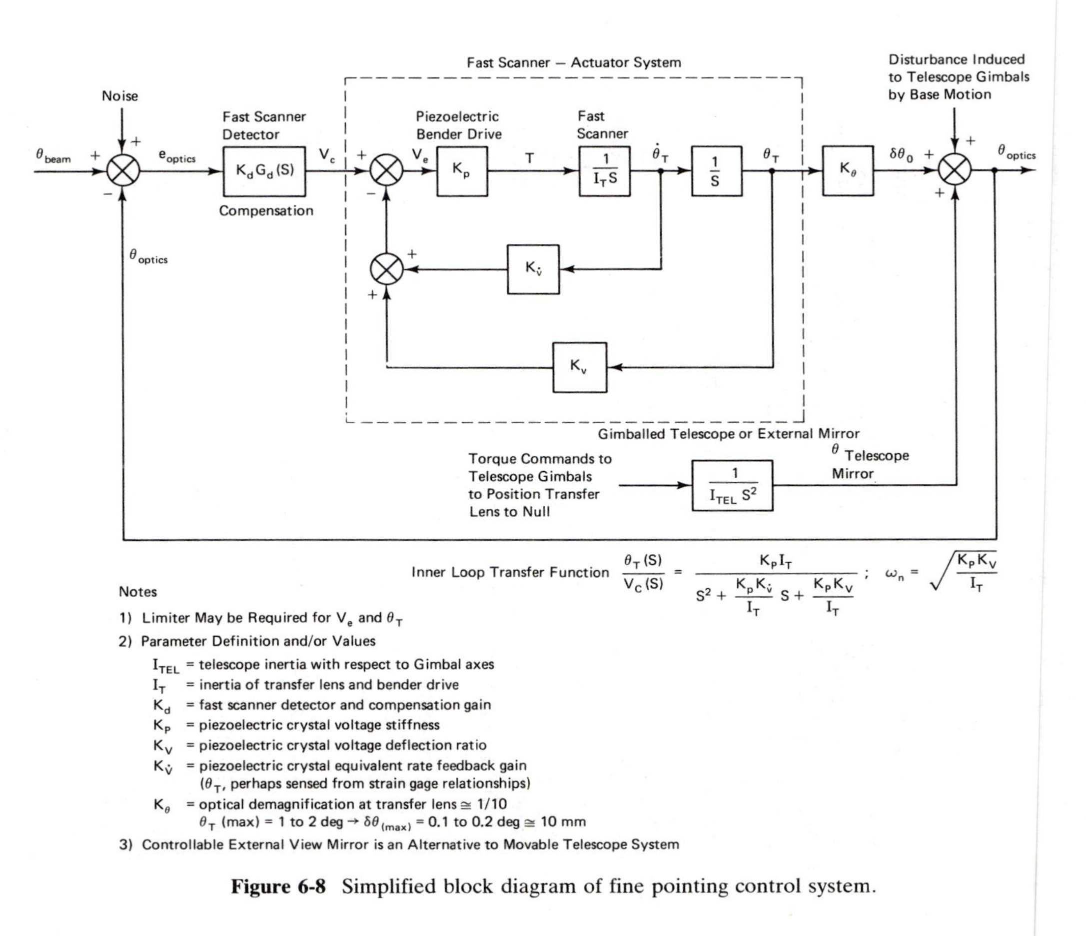

Figure 6-8 Simplified block diagram of fine pointing control system.

limited, a pointing accuracy of 1 to 2 μrad is required with a desired optical pointing error, e_{optics}, limit of less than 1 μrad.

The fast scanner drive mechanisms produced to date have linear ranges (view angle) as large as $\pm 1.5°$. However, the image falling on the fast scanner is demagnified by the ratio of the primary mirror diameter to the subreflector diameter, typically 10 to 1, and the beam-steering capability of the fast scanner is demagnified by the same ratio. The effective linear range of the fast scanner ($\delta\theta_0$) is, therefore, on the order of 0.1°. Resonant frequencies of 100 to 500 Hz are possible for the lens-actuator mount.

From the above constraints, a minimum loop gain (K_{loop}) can be determined which would be required to keep $e_{optics} < 1$ μrad when $\delta\theta$ is approximately 3000 μrad. For this condition K_{loop} equals to $(\delta\theta_0/e_{optics}) > 3{,}000$. Alternatively, this can be used to determine, as a function of frequency, the maximum attitude disturbance which could be induced to the telescope by base motion (satellite, aircraft, etc.) and still have the fine pointing control system maintain $e_{optics} < 1$ μrad.

Figure 6-9 illustrates the maximum disturbance motion the fast scanner actuator system can attenuate. In other words, the curves represent the maximum allowable base motion, as a function of frequency, which is acceptable and still be able to maintain optical pointing errors <1 μrad. A simplified type 1 control system (integrator) was assumed in defining the curves for the frequency range of interest. Different families of curves are given to account for variations in the fast scanner system bandwidth. This data can be used to estimate the compatibility of the base vehicle and its own attitude control system with that of the telescope fine pointing system. A most desirable feature is to have the telescope fine pointing system adaptable to all base vehicles which may be candidates in a laser space communication network.

6.4.3 Critical Control Design Areas

Basically, there are three critical areas of major importance in evaluating the fine pointing system fast scanner lens-actuator loop. They are: the desired pointing accuracy, the effective linear travel of the scanner mirror, and the resonant frequency of the actuator. These areas determine the system bandwidth of the transfer crystal device.

The desired pointing error limit of 1 μrad proposed is approximately 10 percent of the beamwidth with the telescope operating near the diffraction limit. This value was used only in determining base mount motion attenuation capability of the fine pointing system and does not include additional uncertainties or errors within the telescope caused by elements used in mechanical alignments and/or sensing devices. These errors would be statistically combined with the pointing error induced by base motion. A reduction in this

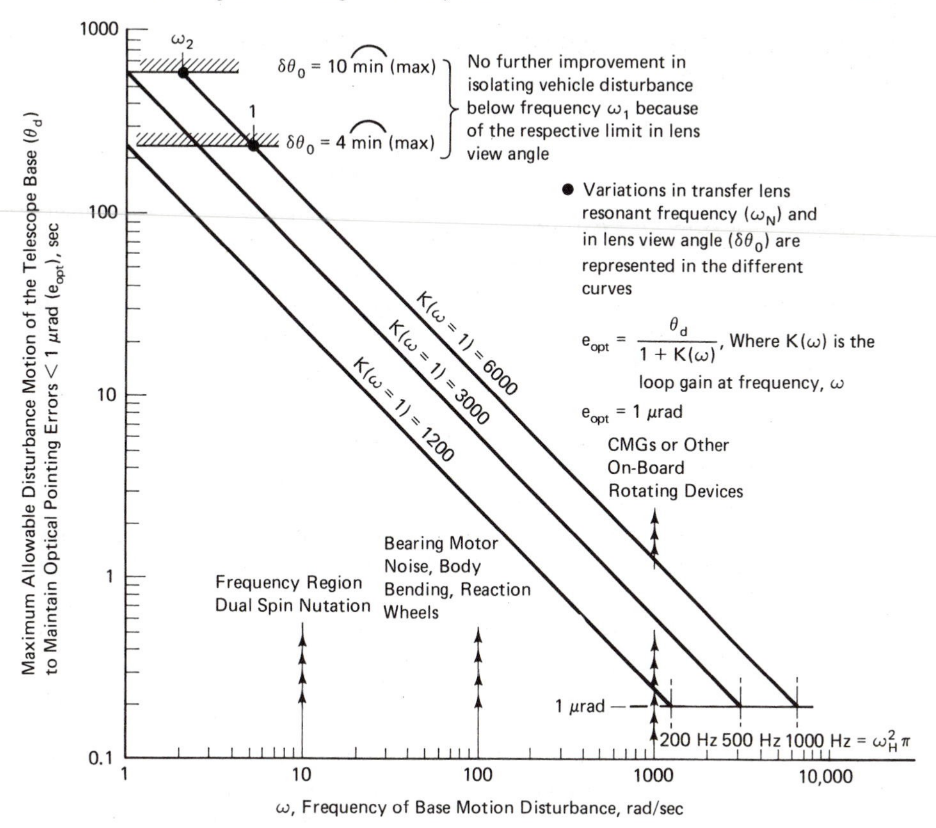

Figure 6-9 Maximum wideband isolation of optical axis achievable with piezoelectric fast scanner actuator.

limit (increasing the acceptable error limit) would help alleviate the other two critical design areas or could be used to reduce the loop gain requirements.

The effective linear range of the fast scanner and the resonant frequency of the piezoelectric bender drive tend to be interrelated. In general, an increase in piezoelectric bender drive resonant frequency causes a reduction in the linear range of travel of the fast scanner.

Also of importance in the use of optical beam deflectors are methods for determining lens position or deflection and its rate. For the piezoelectric devices the deflection angle is not proportional to the applied signal because of hysteresis and other nonlinearities inherent in the materials. Some work reported in this area has utilized strain gauges, polarization measurement techniques, and differential photocells for deflection sensing. The need for accurate determination of the lens position is required if fine pointing accuracy is to be achieved.

6.5 LASERCOM ATP COMPUTER SIMULATION PROGRAM DESCRIPTION

A digital computer simulation program can be extremely useful for verifying a subsystem design. Such a program (called LASCON) has been developed by The Aerospace Corporation which can be used to evaluate system concept feasibility, validate the required performance, and identify the sensitive parameters in the lasercom acquisition, tracking and pointing control system for various future communication satellites. The program is written in FORTRAN and resides on the Aerospace CDC CYBER computer. The main features of the program are:

1. Modular program structure for various satellite configurations.
   Since various future communication satellites may have different configurations, attitude stabilization and determination systems, and operating environments, the LASCON program is written in modular structure providing flexibility to replace modules for each different design. Moreover, the modular structure will also provide subsystem performance evaluation (e.g., coarse gimbal dynamics and control servo analysis) before the total simulation program is completed.
2. Detail cross-coupling between the lasercom subsystem and the spacecraft.
   There are control systems in the spacecraft as well as in the lasercom subsystem. It is inevitable that cross-coupling may exist between the various control devices, the spacecraft, and the lasercom supporting structure. A completely coupled dynamic formulation is used which integrates all the state variables simultaneously with variable step size to provide the capability of investigating possible cross-coupling effects within the entire system.
3. High-fidelity component modeling for parameter sensitivity analysis.
   The components of the lasercom ATP subsystem (such as the gimbal bearings, tracking detectors, fine-steering mirrors, etc.) are modeled with high fidelity to provide the capability for parameter sensitivity studies. The purpose of a sensitivity study is to identify critical components and risk in component technology development. The component models should be updated as laboratory brassboard testing data becomes available. Moreover, it can also be used for determining realistic error budget allocations so appropriate specifications can be defined.
4. Profile of pointing error time history for mission sequence evaluation.
   The final output of the LASCON program is to provide the laser beam line of sight vector pointing error during each mission time sequence under various spacecraft dynamic conditions. Thus, as indicated in Figure 6-10, a reference path is established which generates the ideal states

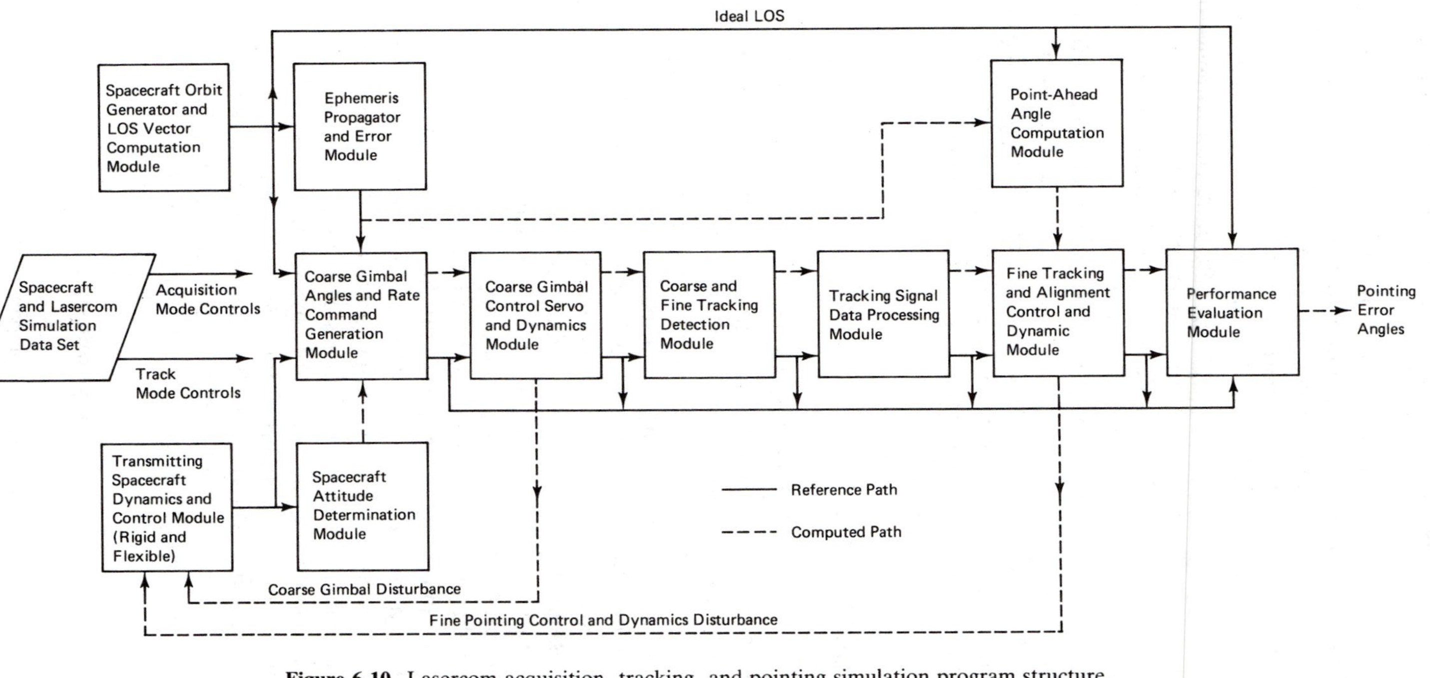

Figure 6-10 Lasercom acquisition, tracking, and pointing simulation program structure.

for the spacecraft attitude, the transmitting and receiving laser beams, and the ephemerides of the communication link. As the computation procedure flows through each functional module, this reference path can be contaminated by a particular module and local performance evaluation of that specific module becomes possible. The time history output plots will enable the user to evaluate mission sequence performance, especially in the acquisition phase of the mission.

5. Acquisition and tracking mode transition via master table lookup control.

   A master table look-up time sequence and mode control flags resides in the LASCON master input data set to control the open loop acquisition mode functions, handover, and the closed loop tracking and pointing mode functions. Individual assessment of either the acquisition sequence or the tracking and pointing feedback loop functions is possible via the input command with no interrupts.

6.6 CONCLUSIONS

The tracking and precision pointing control of a laser communication system for future space applications has been presented in this chapter. The 1 $\mu\Omega$ pointing, accuracy and submicroradian jitter stability requirements for closed loop tracking and pointing, together with requirements for rapid and near autonomous command and control during initial acquisition and communication, present some unique problems in the area of spacecraft control. In this chapter, system specifications and components needed for a future space lasercom system are outlined. The feasibility of developing a common control subsystem for various lasercom designs is also discussed and appears possible for various future space communication applications.

An evaluation of the acquisition and tracking subsystem in terms of requirements and performance capability has been presented for a typical design application. A discussion of performance sensitivity, for these subsystems, to critical components or design issues has also been given. It should be noted that since a lasercom system in space has not yet been demonstrated, design areas of concern that were not covered may evolve. The purpose of establishing a detailed computer simulation program is to identify some of these potentially uncovered areas. It is vitally important that simulation and analysis activities be worked in conjunction with laboratory brassboard testing to ensure the success of a space-based Lasercom program.

REFERENCES

1. D. Botez, and G. T. Herskowitz, "Components for Optical Communication Systems: A Review," *Proceedings IEEE,* Vol. 68, No. 6, June 1980.

2. H. M. Stoll, "Optimally Coupled, GaAs—Distributed Bragg Reflection Lasers," *IEEE Transition on Circuits and Systems*, Vol. CAS-26, No. 12, Dec. 1979.
3. R. Gagliardi and S. Karp, *Optical Communications* (New York: John Wiley and Sons, 1975.)
4. Pratt, *Laser Communications* (New York: John Wiley and Sons, 1970).
5. M. Ross, J. Brand, and G. Lee, "Short Pulse Laser Modulation Techniques," AFAL-TR-70-130, McDonnell Douglas Astronautics Company, St. Louis, Missouri, June 1970.
6. D. G. Aviv, "Summary of Proposed Advanced Development Program for Laser Space Communication Subsystems," TOR-0066(5304)-2, March 1970, The Aerospace Corporation, El Segundo, CA.
7. J. M. Lopez, "Deep Space Laser Communication Study—Acquisition and Tracking Subsystems Evaluation," TOR-0054(6312-02)-2, Jan. 1971, The Aerospace Corporation, El Segundo, CA.
8. E. S. Clarke, and H. D. Brixey, "Acquisition and Tracking System for a Ground Based Laser Communication Receiver Terminal," *25th SPIE's International Symposium*, August 24, 1981.
9. K. Yong, C. C. Chao, and A. S. Liu, "Autonomous Navigation for Satellites Using Lasercom System," *AIAA/AAS 3rd Winter Conference*, Jan. 1983.

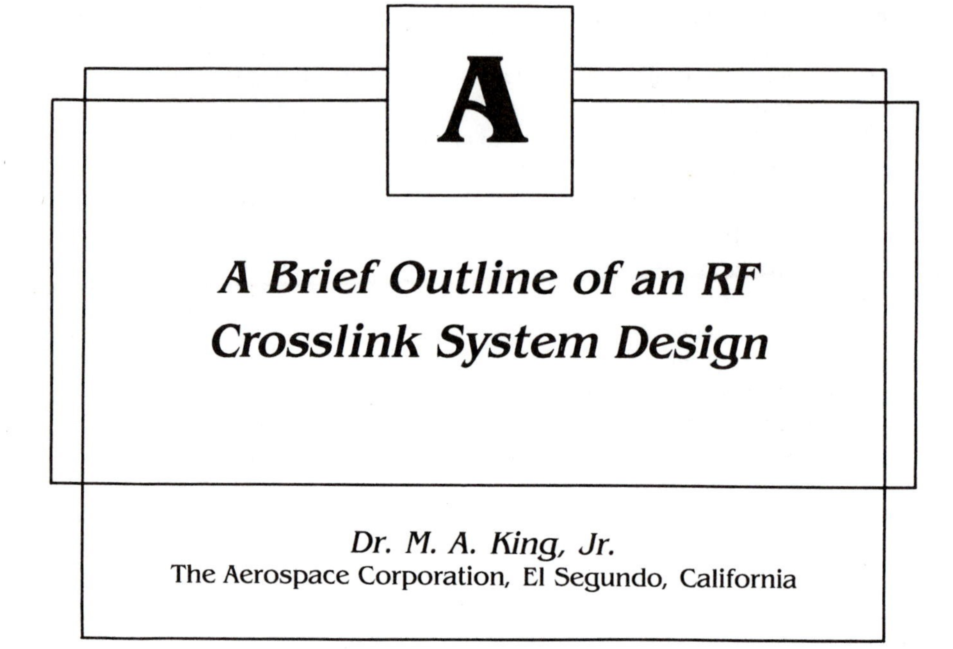

A
A Brief Outline of an RF Crosslink System Design

Dr. M. A. King, Jr.
The Aerospace Corporation, El Segundo, California

A.1 INTRODUCTION

This appendix is a brief outline of the major features in a radio frequency (RF) communications system design. The intent is to provide a counterpoint to the optical system designs discussed in the main text. The development will concentrate on RF satellite crosslinks whose carrier frequencies are in the vicinity of 60 GHz. The concentration on crosslinks is in keeping with the orientation of the text because crosslinks are often easier to analyze than uplinks, downlinks, or terrestrial links. This is because a crosslink design need not deal with the complex interactions of the RF signal with various atmospheric phenomena. The carrier frequency was selected because it is in one of the frequency bands assigned to crosslink usage by international agreement (WARC 79), and because of spectral design considerations. The design considerations center around the fact that many oxygen absorption bands lie in the vicinity of 60 GHz. This means that a satellite crosslink system using these bands will not have to deal with the volume of stray RF radiation from terrestrial sources that plague some of the more heavily used frequency bands.

The most important design tool for any RF link design is the link equation. This equation allows the trade-offs between major link design parameters to be easily and quickly accomplished. The link equation is an expression of the signal-to-noise ratio (SNR), usually referenced to the receiver, in terms of the various gains and losses in the transmitter, the receiver, and in the transmission path. Since these gains and losses are typically multiplicative, the link equation is most often expressed in decibels (dB). This allows the mathematics to be done as additions and subtractions instead of multiplications and divisions.

A number expressed in decibels is equal to ten times the logarithm of the number to the base ten—that is,

$$X(\mathrm{dB}) = 10 \log_{10}(X)$$

The discussion in this appendix will concentrate on the development and use of the link equation for RF crosslink system design.

Both RF and optical crosslink systems have inherent strengths and weaknesses. This being true, the best system solution depends on the context. System features that could lead to a significant advantage of one system over the other will be pointed out in the discusssion that follows. It will be seen that the relative advantages are typically found in the weights of the antenna systems that are required in order to meet system data rate and error probability specifications. For a given error probability requirement, RF systems usually have an advantage at low and medium data rates, while optical systems often become more advantageous at very high data rates. These ideas will be developed through the use of a link equation analysis.

A.2 THE LINK EQUATION

The link equation will be developed in terms of the three major link components, the transmitter, the channel, and the receiver. A simplified block diagram of the functions within the transmitter and receiver is given as Figure A-1. Associated with the transmitter is the output transmitter power. This is determined by the power amplifier output and the transmit antenna gain. Associated with the channel are most of the transmission loss terms. For crosslinks the most significant loss term is the free space loss—the reduction in signal energy density caused by the geometric dispersion of the signal energy with distance from the transmitter. Associated with the receiver are a mixture of gains and losses. The gains are mostly associated with the receiver antenna. The losses are related to system noise which is typically associated with the receiver.

In symbols, the link equation may be expressed as

$$(C/N_o) = (EIRP) - L_T + [G_r - (kT)] \quad (\mathrm{dB\text{-}Hz}), \qquad (\text{A-1})$$

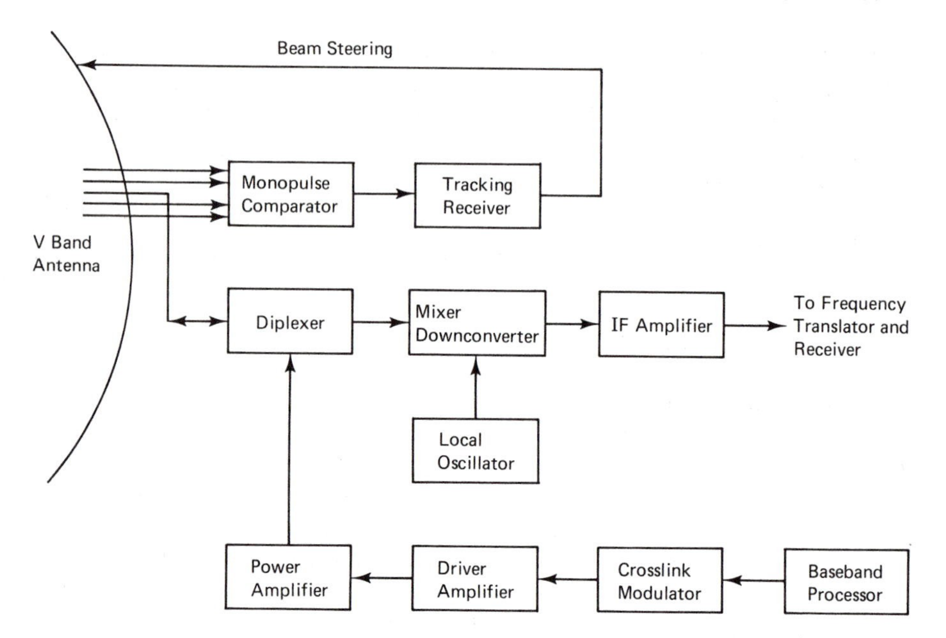

Figure A-1 60-GHz crosslink transmitter and receiver simplified block diagram.

where

(C/N_o) = received carrier-to-noise energy density ratio (dB-Hz).
$(EIRP)$ = transmitted *E*ffective *I*sotropic *R*adiated *P*ower (dB)
L_T = total channel losses (dB)
G_r = receiver antenna gain (dB)
(kT) = effective receiver noise energy density (dB-Hz)

The receiver noise energy density is more often expressed in its component parts where k = Boltzmann's constant = -228.6 (dB-Hz/K) and T = effective system noise temperature (dB-K).

A significant difference between RF and optical links can be seen in Equation (A-1). This difference is that for RF systems the major noise source is added to the signal in the receiver, while in many optical systems, the major noise source is an inherent part of the signal itself. This is a consequence of the fact that the lower energy RF photons and the form of RF detector technology make photon limited operation at RF impractical. The major noise source in the typical satellite RF link is the thermally induced shot noise, or Johnson noise, in the RF detector circuitry. Thus, the noise in the typical satellite RF link is independent of the signal and is well approximated as an additive white Gaussian noise, over most spectral ranges of interest.

A.2.1 The Transmitter

A commonly used figure of merit for the transmitter is Effective Isotropic Radiated Power (*EIRP*). In symbols, an expression for this quantity is

$$(EIRP) = P_t - L_l + G_t \quad (dB) \tag{A-2}$$

where

P_t = transmitter power
L_l = line and coupling losses
G_t = transmitter and antenna gain

Transmitter Power

The transmitter power, P_t, in Equation (A-2) is the power available to the individual crosslink channel. This is as opposed to the maximum power that the transmitter could theoretically generate. If the link under consideration is comprised of several channels that are separated in frequency (an option not typically available to optical systems), the total output power will usually be reduced to ensure that the output power amplifier operates in its linear region. A disadvantage in using multiple channels separated in frequency is that if the transmitter is operated at full power, the nonlinear saturated gain characteristic will cause the different channels to interfere with each other. To avoid this problem, the transmitter is "backed-off" from its maximum possible power output into the amplifier's linear region. Let this amount of back-off be denoted B_o(dB). Let the fraction of the remaining power that is to be assigned to the channel of interest be denoted by γ(dB). Then if P_{MAX} denotes the maximum total output power deliverable by the power amplifier,

$$P_t = P_{MAX} - B_o + \gamma \text{ (dB)} \tag{A-3}$$

The maximum transmitter power, as is true in optical systems, is an important design parameter. The transmitter power can play a major role in the determination of the size, weight, power requirements, and cost of the satellite. A benefit of a link equation as a design tool is that it allows the trade-offs between these link design parameters to be easily accomplished. Such a trade-off will be developed later.

The technology available for 60-GHz power amplifiers is split into two areas. Dominating the high power region are the traveling wave tube amplifiers (TWTAs). The state-of-the-art at 60 GHz is about 5 W.[1] However, there are current design efforts intended to produce tubes of considerably greater output. Figure A-2 presents an estimate of potential tube development through

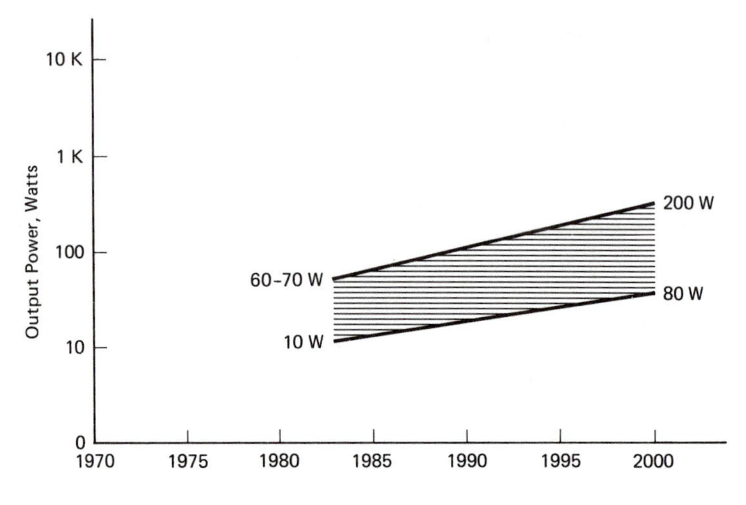

Figure A-2 60-GHz TWTA projections (coupled cavity).

the end of the century. It will be seen from the figure that the maximum output TWTAs hypothesized for the near future produce an output power in the vicinity of 100 W.

Dominating the lower power applications are IMPATT (Impact Avalanche and Transit Time) diode amplifiers. The state-of-the-art for IMPATT amplifiers is about 1.5 W, but 5 W may be achieveable.[1] In addition, it may be possible to coherently combine the outputs of several low power devices in order to produce a high reliability, space qualified device generating 10 to 40 W by 1986.[1] IMPATT diodes have the advantages of greater inherent reliability and smaller size and weight, relative to TWTAs. If coherent power combining proves feasible, TWTAs may be relegated to only the highest power requirement applications. A projection of IMPATT device output power by both Hughes Aircraft Company and the Raytheon Corp. indicates that in 1986, about 1.5 W devices will be available at 60 GHz.

Line Loss

The exact value for line and coupling loss, L_l, will depend upon a spacecraft's physical design. For a three-axis stabilized satellite, where the power amplifier is mounted near the antenna, the loss will be small. For a spin stabilized satellite, where the power amplifier is connected to the antenna by rotary joints and lengths of wave guide, the loss may be considerably more. For the purposes of this outline, a value of 1 dB will be assumed. This number is probably conservative (high) for a three-axis stabilized satellite, but may be optimistic for a spin stabilized design.

Transmitter Antenna Gain

An expression for boresight antenna gain, G_t, is given by:

$$G_t = 10 \log(4\pi A\eta_a/\lambda^2) \text{ (dB)}, \tag{A-4}$$

where

A = aperture area
η_a = antenna efficiency
λ = transmission wavelength

For a circular aperture, a typical value of antenna efficiency of 55 percent, and the transmission frequency of 60 GHz, Equation (A-4) can be rewritten as:

$$G_t = 43.0 + 20 \log(D) \text{ (dB)}, \tag{A-5}$$

where D is the aperture diameter in feet.

A.2.2 The Channel

The number and identity of the terms describing a communication channel is more a matter of agreement than of standardized definition. For the purposes of this outline, the channel will include all of the signal loss terms that are located between the transmitting and the receiving antennas. For crosslink systems, the most significant loss term is the free space loss. This is the loss, relative to the receiver, of signal energy due to the transmitted energy's geometric dispersion with distance. The expression for free space loss is

$$L_f = 20 \log(4\pi r/\lambda) \text{ (dB)}, \tag{A-6}$$

where r = intersatellite distance and λ = transmission wavelength. For a transmission frequency of 60 GHz, this loss can be expressed as

$$L_f = 128.0 + 20 \log(r) \text{ (dB)} \tag{A-7}$$

where r is expressed in kilometers. For two satellites at synchronous altitude, operating such that the line-of-sight between them is effectively above the atmosphere (>75 mi), the maximum intersatellite range will be approximately 81,500 km. Thus, for synchronous satellites operating at 60 GHz, the maximum space loss will be approximately 226.2 dB.

Another loss that should be considered in the design of a crosslink channel would be antenna pointing and polarization loss, L_a. If the transmissions are circularly polarized, the polarization loss is probably negligible.

The pointing loss however, depends on the antenna tracking accuracy. An approximate expression for pointing loss is[2,3]

$$L_a = 12(\Delta\theta D/70\ \lambda)^2\ (\text{dB}) \tag{A-8}$$

where $\Delta\theta$ is the angular tracking accuracy. If spacecraft dynamics can be ignored, as may be the case for synchronous satellites, a closed loop antenna tracking system can maintain accuracy to about a tenth of the antenna half power beam width[1] ($70\lambda/D$). Thus, this loss, including both transmit and receive antennas, will be overbounded by 0.5 dB. In cases where spacecraft dynamics cannot be ignored, this loss can, of course, be large. If another 1 dB is added in to account for crosstalk and miscellaneous implementation errors and degradations, the total channel losses for a 60-GHz link between two synchronous satellites will be

$$L_l = 227.7\ \text{dB}$$

A.2.3 The Receiver

The receiver is portrayed in Equation (A-1) as being described by two major design parameters. These are the receiver antenna gain and the system noise density. For most satellite crosslink systems (see Figure A-1), a single antenna will be made to serve both the transmitter and the receiver. Therefore, ignoring minor differences in the transmit and receive frequencies (minor relative to 60 GHz), from Equation (A-5) we can write

$$G_r = G_t = 43.0 + 20\log(D),\ (\text{dB}) \tag{A-9}$$

The system noise density is the product of Boltzmann's constant and the effective system noise temperature. Expressed in decibels, Boltzmann's constant, k, is -228.6 (dB-Hz/K). The effective system noise temperature is given by Ref. 3 to be

$$\tau = T_a/L + (L - 1)\tau_o/L + T_r\ (\text{K}) \tag{A-10}$$

where

T_a = antenna noise temperature
L = line and coupling losses
τ_o = ambient temperature
T_r = noise temperature of the low noise receiver

The antenna noise temperature is determined by the radiating objects in the antenna beam pattern. For a 60-GHz crosslink, the antenna pattern will "see" mostly cold space. Therefore, the antenna noise temperature will probably be less than 10K. The line and coupling losses can be assumed to be the same as for the transmitter—i.e., 1 dB = 1.26. The ambient temperature in the vicinity of the crosslink receiver depends upon the spacecraft design, but it could be expected to be 300K at most. Low noise receivers operating at 60 GHz have been built under laboratory conditions that have exhibited effective noise temperatures as low as 1000K.[4] However, for a space qualified application 3,000K appears to be a more reasonable value and it is unlikely that temperatures much lower than 2,000K could be hoped for.[1] Thus, the effective system temperature is

$$T = 10 + (0.26)(300)/(1.26) + 3{,}000 = 3{,}072\text{K} = 34.9 \text{ (dB-K)}.$$

It is common for the receiver gain and effective noise temperature to be combined into a single receiver figure of merit. Under the assumptions at hand, this figure of merit is

$$\begin{aligned}(G_r/T) &= [43.0 + 20\log(D)] - 34.8 \\ &= 8.2 + 20\log(D)\end{aligned} \qquad \text{(A-11)}$$

where the antenna diameter, D, is in feet.

A.3 LINK EQUATION—INTERMEDIATE RESULTS

The results obtained thus far will now be substituted into Equation (A-1). Thus,

$$\begin{aligned}(C/N_o) &= (EIRP) - L_T + [G_r - (kT)] \quad \text{(dB-Hz)} \\ &= [P_t - 1.0 + 43.0 + 20\log(D)] - 227.7 \\ &\quad + [43.0 + 20\log(D) + 228.6 - 34.9] \\ &= P_t + 40\log(D) + 51.1 \quad \text{(dB-Hz)}\end{aligned} \qquad \text{(A-12)}$$

Equation (A-12) presents an expression for carrier to noise spectral density ratio for a 60-GHz crosslink that is operating at an intersatellite range of 81,500 km. The expression is in terms of the available transmitter power and the antenna diameter. If a requirement on the minimum acceptable (C/N_o) is established, Equation (A-12) can be used to make system trade-offs between satellite power and antenna size. This kind of trade-off will be developed below.

A.3.1 Carrier-to-Noise Ratio Requirement

For digital communications systems, the requirement on (C/N_o) is usually expressed in terms of the required energy per information bit to noise energy density ratio (E_b/N_o) that is required to obtain some specified bit error probability. The relationship between (C/N_o) and (E_b/N_o) is given by

$$(C/N_o) = (E_b/N_o) + R + M \qquad \text{(A-13)}$$

where

R = information bit rate
M = system link margin

For the case of a link between a satellite and an earth station, the margin, M, allows for variations in the transmission characteristics of the atmosphere, and the effects of these variations upon signal strength. For a crosslink, however, the margin would mainly provide an additional degree of safety against unanticipated system degradations. Typically, the margin for a crosslink will be small—one or two dB.

A.3.2 Bit Error Probability Requirement

The required value for (E_b/N_o) depends on the required bit error probability, the modulation scheme, and the amount of signal processing that is used. The requirements on error probability will vary with the use to which the data will be put. The variations in error probability requirements can be from as high as 10^{-2} to 10^{-3} for digitized voice signals to as low as 10^{-10} to 10^{-12} for computer data records. The value that will be used in this outline will be 10^{-5}. This is an error probability of fairly general applicability.

A.3.3 Modulation Schemes

Modulation schemes represent an area of major difference between RF and optical communication systems. Optical systems generally use direct energy detection. This tends to limit optical systems to modulation schemes that are variations of on-off keying (OOK). One example of such a variation is pulse position modulation (PPM). Radio frequency systems, on the other hand, generally use heterodyne receivers, and use modulation techniques that make use of this advantage. There are two substantial advantages to heterodyne operation. The first is that it allows the transmission of a constant energy signal. Constant energy signalling is typically easier and more energy efficient than pulsed signalling. The second advantage is that heterodyne operation allows the use of signal parameters other than energy presence or absence to carry the information. The two parameters commonly used are signal phase and signal frequency. The advantage of using auxillary parameters is best seen in an example. The detection of a binary signal with a heterodyne receiver involves a decision between two possible signals with known characteristics. The detection algorithm utilizes a detection threshold that plays one possible signal off against the second. Since any additive system noise will effect both possibilities, changes in the average noise level will not effect the detector threshold setting. With direct energy detection techniques, however, the decision is between the presence of signal energy or a residual noise background. In this case, the optimum detector threshold setting is critically dependent on accurate knowledge of the average noise level. Thus, a heterodyne receiver enjoys an important and sometimes critical advantage over a direct energy detection receiver. This advantage is that for a heterodyne receiver, the form of the optimal detector is independent of the values of the noise statistics, while the optimal detector is critically dependent on these values with a direct detection receiver.

The form of a typical RF signal can be represented by

$$S(t) = A\cos(\omega_c t + \phi(t)) \qquad \text{(A-14)}$$

where

A = signal amplitude
ω_c = carrier radian frequency
$\phi(t)$ = the information bearing signal function

The exact form of the signal function, $\phi(t)$, depends upon the choice of modulation technique. The two most popular digital modulation schemes are Phase Shift Keying (PSK) and Frequency Shift Keying (FSK). Many other modulation schemes are possible, some involving a time varying amplitude, A, but these others are rarely used on modern digital satellite channels.

A major design decision for digital modulation schemes is the signalling alphabet size. This is the number of different signals that the transmitter can use, and the number of possibilities that the receiver has to consider while detecting the signal. Because of the nature of digital technology, this number is almost always a power of two. The most commonly used alphabet sizes are 2 and 4 for PSK, and 2 and 8 for FSK. The design decision involves a tradeoff between relative speed of information transfer (more bits per symbol with larger alphabets) versus increases in complexity, size, and weight, as alphabet sizes increase.

For an M-ary signalling alphabet, and PSK modulation, the signal function will have the form

$$\phi(t) = \sum_j \gamma_j(2\pi/M)(U(t-(j-1)T) - U(t-jT)), \quad -\infty \le j \le \infty \qquad \text{(A-15)}$$

where

γ_j = jth symbol, $\gamma_j\ \varepsilon(0, 1, \ldots, M-1)$
$U(\cdot)$ = unit step function
T = symbol interval (seconds)

If the signal is FSK modulated, the information function will have the form

$$\phi(t) = \sum_j (\gamma_j - M/2)\Delta\omega t(U(t - (j - 1)T) - U(t - jT)) + \phi_o, \quad -\infty \leq j \leq \infty \qquad \text{(A-16)}$$

where

$\Delta\omega$ = radian frequency signal separation
ϕ_o = arbitrary constant phase shift

Typically FSK signals will be separated in frequency by an integer multiple of the transmitted symbol rate. That is,

$$\Delta\omega = n(2\pi/T),\ n = 1,2,\ldots \qquad \text{(A-17)}$$

This separation provides orthogonal signals in the frequency domain.

Each modulation scheme has its inherent advantages and disadvantages. Observation of Equation (A-15) indicates that for PSK to be effective, the receiver must have accurate knowledge of the phase of the incoming signal. This means that the receiver must contain a phase tracking loop that has the ability to maintain a stable phase reference in spite of the possible extraneous noise or motion induced changes in signal phase. This adds power, cost, and complexity to the receiver. The FSK scheme, on the other hand, is indifferent to the exact value of signal phase. This was emphasized by the inclusion of an arbitrary phase parameter, ϕ_o, in Equation (A-16). The indifference to phase means that a receiver using FSK will probably be simpler than one using PSK for the same alphabet size. Furthermore, the indifference to phase means that FSK will be more or less uneffected by random channel-induced phase changes. The disadvantages of FSK are in relative error performance and bandwidth. These and other differences are developed below.

Performance curves for binary, quaternary, and octal ($M = 2,4,8$) PSK are given in Figure A-3. These curves show bit error probability as a function on information bit to noise energy density ratio (E_b/N_o). Performance curves for binary and octal FSK are given in Figure A-4. Two things will be noticed immediately with respect to Figures A-3 and A-4. The first is that the performance of binary and quaternary PSK is the same. This is because the factor of two decrease in differential power between symbols is offset by a factor of two increase in the number of bits per symbol. Thus, the performance in terms of bits remains the same. The second thing that will be noticed is that the performance of octal PSK is worse than that of binary PSK, but the performance of octal FSK is better than binary FSK. This is due to a fundamental difference in the way that the information signals are generated. It can be deduced from equation (A-16) that the larger the FSK signalling alphabet (the larger the value for M), the larger the amount of bandwidth

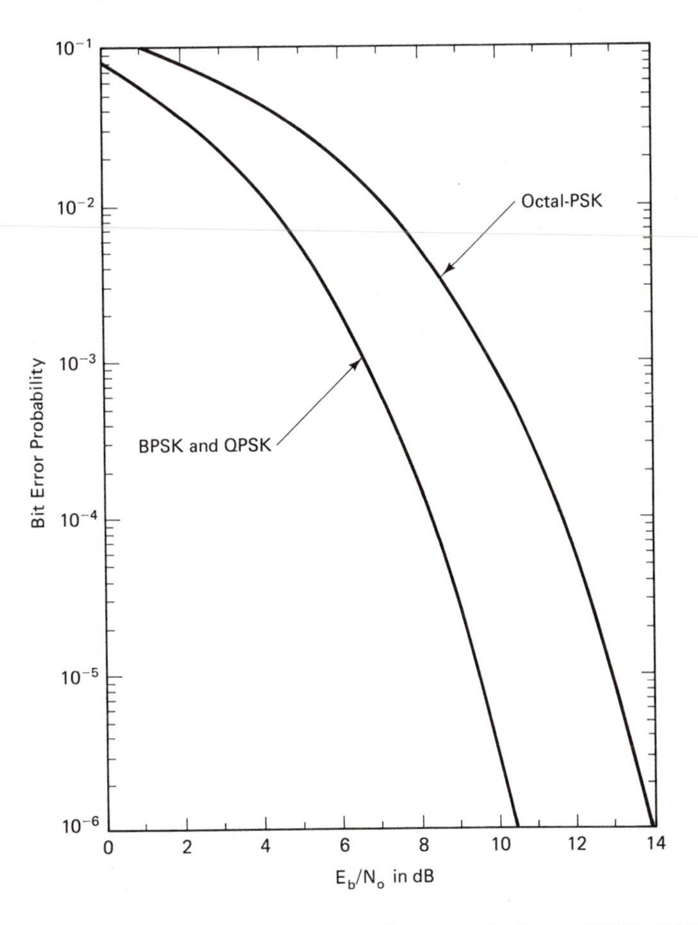

Figure A-3 Bit error probability versus E_b/N_o performance of coherent BPSK, QPSK, and octal-PSK.

the channel will consume. However, other things being equal, the PSK bandwidth is essentially uneffected by alphabet size. Thus, the PSK signals are becoming more crowded, in some sense, while the FSK signals are maintaining their relative distance by spreading out in bandwidth. The fact that FSK signalling inherently requires more bandwidth than PSK, especially for larger signalling alphabets, can make FSK unattractive when bandwidth restrictions are severe.

The curves in Figures A-3 and A-4 are curves of demodulator performance. The performance of the channel can be improved through the use of error correction coding. This is a signalling technique that involves the introduction of a controlled amount of redundancy into the channel. Knowledge of this redundancy can then be used by the receiver to improve the overall link performance.

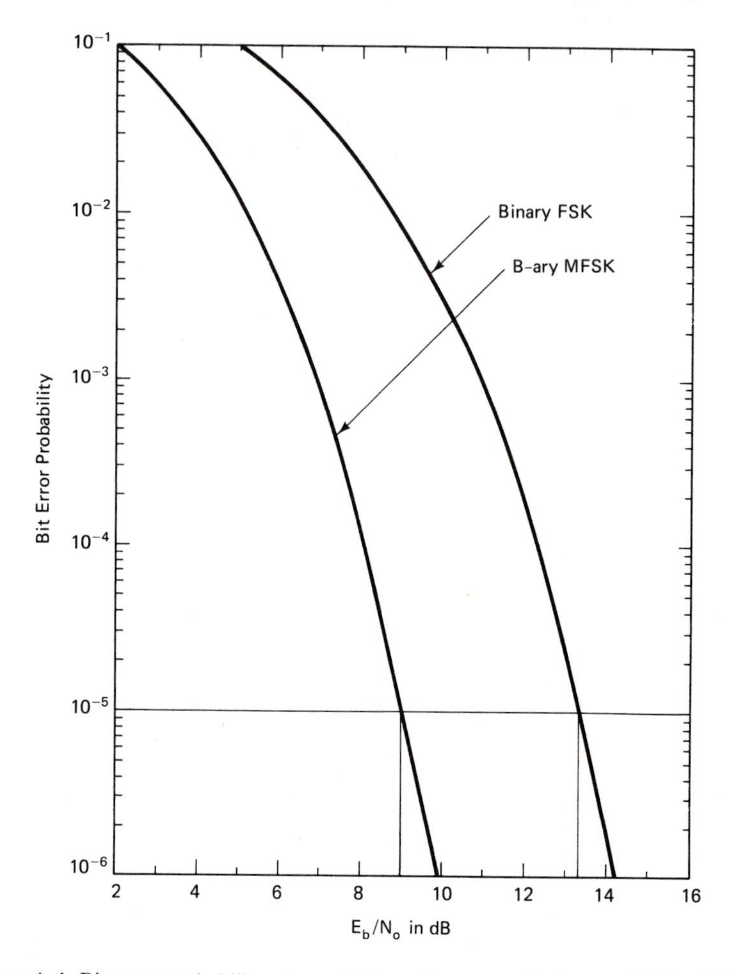

Figure A-4 Bit error probability versus E_b/N_o performance of coherent BPSK, QPSK, and octal-PSK.

A.3.4 Error Correction Coding

The study of error correction coding is far beyond the scope of this brief outline. However, in order to give some feel for what coding may accomplish, Figure A-5 presents a curve of the coding gain of a constraint length 7, rate 1/2, convolutional code used with binary PSK signalling. The coding gain is the amount by which the bit energy to noise energy density (E_b/N_o) may be decreased, and still meet the required output bit error probability, when this code is used. The constraint length is a measure of code complexity, while the rate is a measure of the amount of added redundancy. Convolutional codes are one type of encoding procedure, and this one particular convolu-

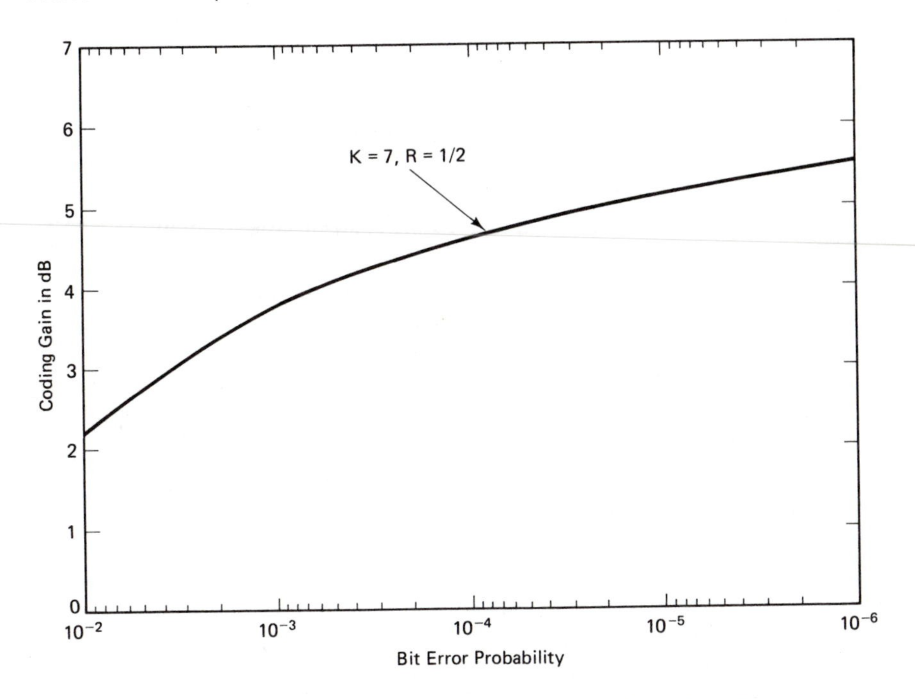

Figure A-5 Coding gain for convolutional error correction coding with BPSK modulation, AWGN, and 3-bit receiver quantization.

tional code has become close to an industry standard for military digital satellite communication channels.

In particular, for a required bit error probability of 10^{-5}, the coding gain is seen in Figure A-5 to be 5.2 dB. The (E_b/N_o) required to produce 10^{-5} bit error probability on an uncoded channel using binary PSK is seen from Figure A-3 to be 9.6 dB. Thus, using this particular error correction code, the received (E_b/N_o) can be as low as 4.4 dB, and the requirement of a bit error probabilty of 10^{-5} would still be met for communications over an additive white Gaussian noise channel.

A.4 LINK EQUATION RECALLED

Returning to the development of the link equation, combining equations (A-12) and (A-13) yields

$$(E_b/N_o) + R + M = P_t + 40 \log(D) + 51.1 \text{ (dB-Hz)} \qquad \text{(A-18)}$$

Assuming the use of binary PSK, a coding gain of 5.2 dB, a required bit error probabilty of 10^{-5}, and a link margin of 2 dB, Equation (A-18) will provide a relationship between transmitter power, P_t, antenna diameter, D, and information rate R.

$$R = P_t + 40\log(D) + 44.7 \text{ (dB-Hz)} \tag{A-19}$$

As would be expected, Equation (A-19) indicates that an increase in the information data rate, for a constant error probability, must be accompanied by an increase in transmitter power, or an increase in antenna diameter, or both. Thus, a tool has been developed that allows trade-offs between information rate, power, and antenna size to proceed in an easy manner.

Figure A-6 is a graph of antenna diameter versus information rate for three values of transmitter power. The transmitter power values roughly represent a current solid state amplifier (1 W), a solid state combined amplifier or TWTA (10 W), and a value representing TWTA projections for future developments (100 W). It can be seen from this figure that a solid state amplifier and an antenna of a few feet in diameter could provide data rates as high as a few megabits per second. With a large antenna, or the more powerful amplifier, data rates up to several gigabits per second may be obtained. These higher rate systems may not be reasonable design choices, however. Figure A-7[1] presents an estimate of crosslink subsystem weight as a function of data rate for both optical and RF systems. The curves indicate that RF systems should enjoy a clear weight advantage over optical systems for data rates less than about one megabit per second. Between one and ten or twenty megabits per second, neither system has a clear weight advantage. Above several tens of megabits, the weight advantage tends to shift clearly in favor of optical systems. The main reason for this shift in relative weight advantage is the weight of the antenna and its supporting structure. Figure

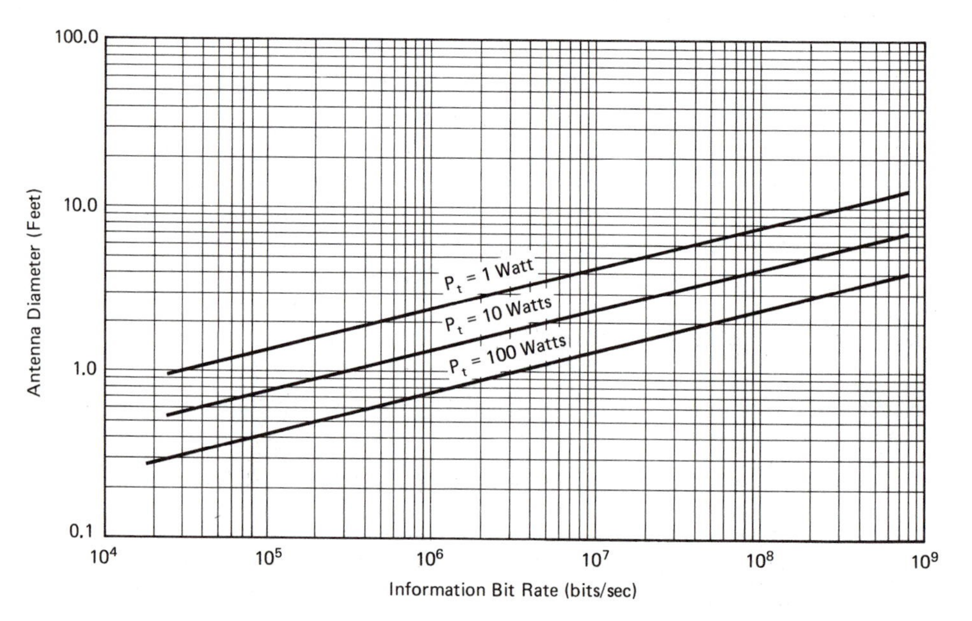

Figure A-6 Information bit rate (bits/s).

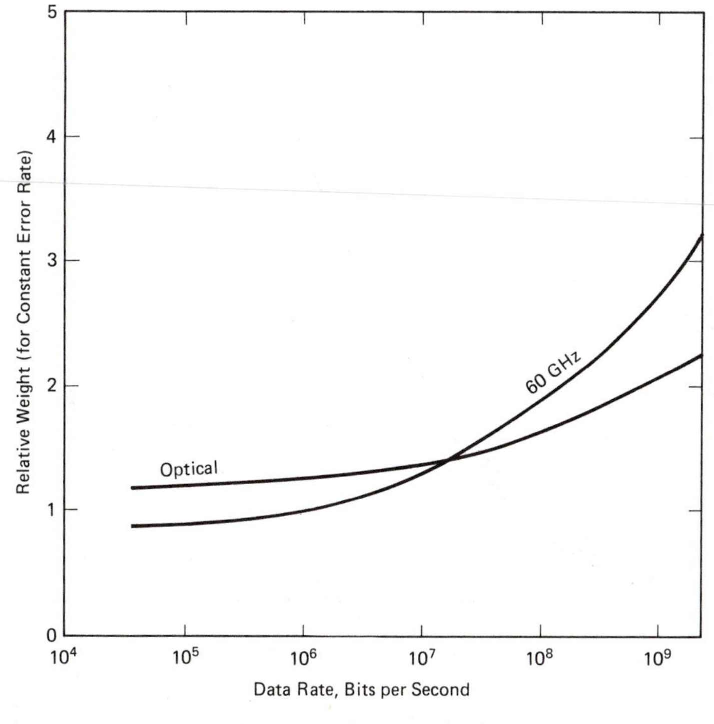

Figure A-7 Data rate (bps).

A-8[4] presents a set of curves of estimated antenna weight as a function of aperture diameter, for three popular antenna configurations. It can be seen that the weight increases vary rapidly as the antenna grows beyond a few feet in diameter, and will quickly become the dominant weight component. While optical antenna systems will also grow with data rate, their smaller original size will allow them a significant advantage at high data rates.

A.5 CONCLUSION

This appendix has presented an outline of the basic design considerations involved in the design of a satellite RF crosslink. Emphasis has been on crosslinks using carrier frequencies in the vicinity of 60 GHz. This was for two main reasons. On the one hand, many systems will consider crosslinks in this frequency band in the future. The band has been assigned for crosslink use by the WARC-79, and the required technologies for the use of these frequencies are becoming available. The second reason was one of ease of analysis. The frequency bands in the vicinity of 60 GHz contain many at-

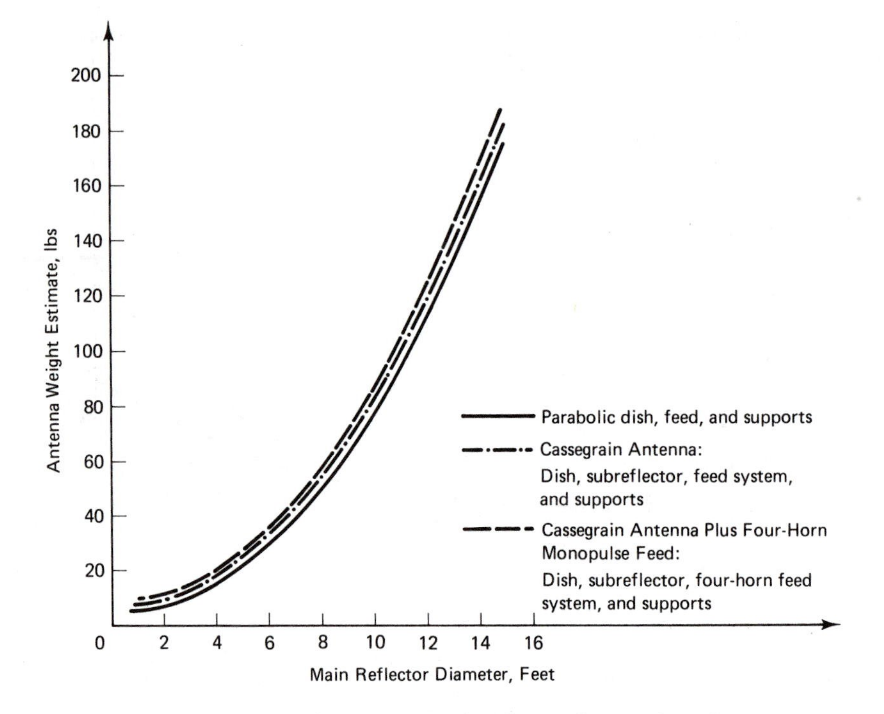

Figure A-8 Antenna weight estimate versus main reflector diameter for reflector-type antenna.

mospheric absorption bands. Thus, by appropriate selection of carrier frequency, a crosslink can avoid the large amount of stray RF radiation emanating from terrestrial sources that plague other frequency bands. The analysis is simplified by the ability to ignore this source of interference and consider only the more classical white Gaussian noise sources.

The main result of this outline is a set of curves of antenna diameter versus data rate for a family of transmitter power values. Implicit in these curves are realistic choices of modulation schemes, amounts of on-board processing, receiver noise levels, and intersatellite ranges. A discussion of the values use is presented in the text. A subsidiary result is the development of the link equation as a simple but very effective design tool for RF communication systems.

The results of the outline lead to a very brief comparison of RF and optical crosslink systems in terms of estimates of subsystem weight versus data rate. The estimates indicate that RF systems enjoy a clear weight advantage at data rates less than about a megabit per second. Optical systems enjoy an increasing weight advantage at data rates above a few tens of megabits per second. In between these limits, neither system has a clear advantage, and design decisions would need to be based on more detailed reasoning.

These results support the opinions stated in the opening paragraphs of this appendix. Both optical and RF systems have their inherent strengths and weaknesses, and the best system for a task depends on the nature of the task.

REFERENCES

1. "Communication Satellite Technology," TOR-0083(3417-03)-2, The Aerospace Corp., July 15, 1983.
2. D. J. Frediani, "Technology Assessment for Future MILSATCOM Systems," Project Report DCA-4, August 24, 1978.
3. J. J. Spilker, *Digital Communications by Satellite* (Englewood Cliffs, N.J.: Prentice-Hall, Inc., 1977).
4. V. K. Agarwal, and D.A., Taggart, "60 GHz Crosslink Technology Study," ATM 79(4417-01)-8, The Aerospace Corp., May 14, 1979.

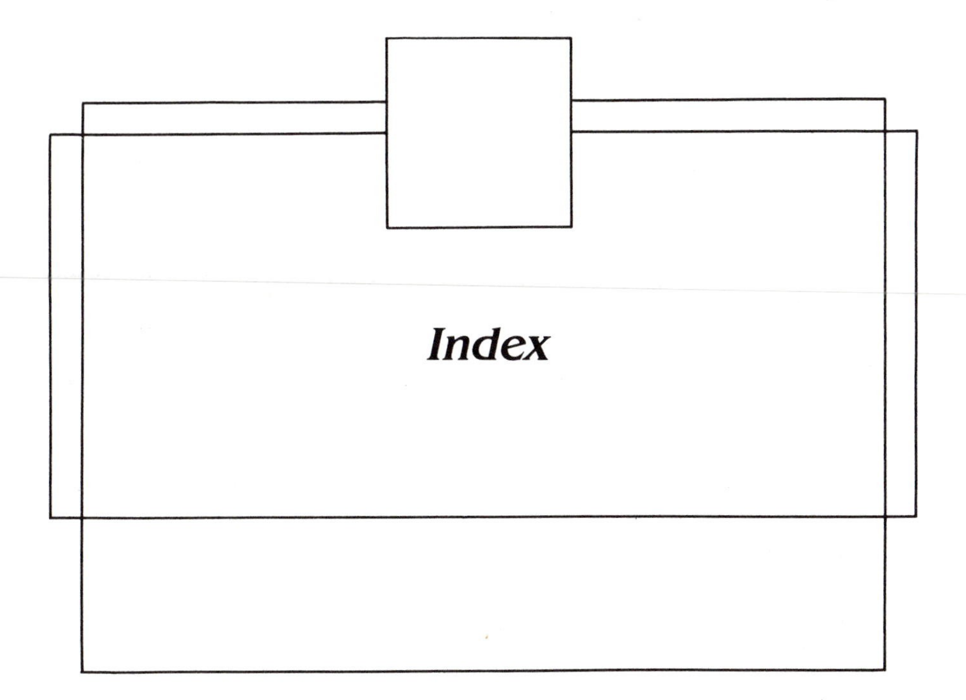

A

B

C

D

E

F

G

H

I

L

M

N

O

P

Q

R

S

T

V

W